The World of
AVIATION

The World of
AVIATION
Chris Ellis

Hamlyn

London · New York · Sydney · Toronto

Acknowledgements

The publishers are grateful to the following individuals and organizations for the illustrations in this book: Air Canada; Air France; American Airlines; Associated Press; Bell Helicopter Company; The Boeing Company; British Aircraft Corporation; British Airways; Austin J. Brown; Charles E. Brown; Neill Bruce; Canadair; Cessna; ECP Armées; Chris Ellis; *Flight*; Fokker; Fox Photos; Goodyear; R. J. Hale; The Hamlyn Group Library; A. C. Harold; David Hodges; Imperial War Museum; Lufthansa; The Mansell Collection; McDonnell Douglas Corporation; Kenneth McDonough; National Army Museum; Novosti Press Agency; North American Aviation; Simon Piercey; Piper; Radio Times Hulton Picture Library; M. Roger-Viollet; Maurice Rowe; The Science Museum; Sikorsky Aircraft; Smithsonian Institution; Nigel Snowdon; Swissair; John W. R. Taylor; *The Times*; United Press; US Air Force; US Army; US Forest Service; US Navy; Westland Aircraft; R. J. Wilson.

629.13
E LL

Published by The Hamlyn Publishing Group Limited
London · New York · Sydney · Toronto
Astronaut House, Feltham, Middlesex, England
Copyright © The Hamlyn Publishing Group Limited 1977

ISBN 0 600 30318 7

Printed in Great Britain
by Jarrold and Sons Limited, Norwich

Contents

Introduction **7**
The Formative Years **8**
War in the Air **36**
The Infant Airlines **60**
The Light Aircraft Era **79**
Helicopters **99**
The Age of Air Power **119**
The Age of the Airlines **168**
Epilogue **190**
Index **191**

Introduction

The world of aviation is an exciting one and the story has never really stopped since ancient times. In the present century it has gained an amazing momentum and even after two world wars, and lots of smaller ones involving aircraft, there is no sign of any slowing up in the pace of development of military aviation. And as airlines shrink the world there is no end to the profusion of new ideas and new aircraft types to capture the public imagination.

This book surveys the story to date as comprehensively as space allows, and I make no apology if the pioneers of the most original ideas get rather more space than others who followed on. I have kept strictly to aircraft, whether lighter or heavier than air, for missiles and spaceships really deserve their own stories. Indeed, some enthusiasts—myself included—might consider them rude intruders in the otherwise well ordered world of true flying machines! Regrettably space limitations in this book do not permit every 'classic' aircraft to be mentioned, and those that are must sometimes have an all too brief coverage. But the broad picture of history is the aim of this book, not a detailed exposition of individual aircraft or flyers.

C.E.

opposite
A McDonnell Douglas F—4K
Phantom of the Fleet Air Arm
being launched by steam
catapult aboard HMS *Ark Royal*

The Formative Years

To thinking men in ancient and medieval times it seemed it *must* be possible–if birds and bats could fly, why not men? The ideas and theories of some of these potential aviators of centuries past have been surprisingly well documented, and range from the ancient Chinese with their paper wings around 2,000 BC to the best known of all, Leonardo da Vinci's man-powered ornithopter which still survives as a plan though it was never actually built. None of the many fanciful ideas–some of them no more than myths or legends–ever resulted in successful flight, though no doubt a good number of intrepid men perished in the attempt to make history as the first airmen many hundreds of years ago.

The urge to fly

It took a long, long time before the nature of the problem was appreciated, let alone overcome. Man's first instinct was literally to imitate the birds and turn himself into a creature which could fly by no other means than by attaching either feathers or flapping wings to his person. In theory, a man so equipped simply had to flap his arms (with the wings attached), and leap into space. Some foolhardy souls still try this sort of flying even today, usually with predictable results. The legend of the flying man is personified in Icarus, who escaped by flying from King Minos in Crete and whose feathered wings were held together with wax, which melted when he flew too near the sun.

What the ancient wing-flapping pioneers did not fully appreciate was the biological difference between a bird or bat and a man. The bird had evolved over millions of years as an efficient, living flying machine. Respiration and heartbeat (typically 800 times a minute) were developed to provide the immense energy required to lift the weight of the creature and provide the muscle power needed to move the sophisticated wings. Every weight-saving measure is used, including hollow bones, and the weight of any bird (non-flyers excepted) is remarkably and unbelievably low relative to its bulk. An adult heron, for instance, weighs only 3 lb, while a wren averages only $2\frac{3}{4}$ drachms–and a drachm is $\frac{1}{16}$ of one ounce. Flying mammals like the bat have similarly evolved to suit them for their mode of existence.

No wonder then, that man could not compete. He had neither the anatomy nor the physiology to generate the muscle power

needed to get his relatively huge weight and bulk off the ground. And he had the added disadvantage of the weight of the materials, be they wings or something more complicated, which were to provide the lift and wherewithal to fly once airborne.

Of all those who considered the flapping wing ideas in the pre-industrial age, Leonardo da Vinci in the fifteenth century remains the most prescient and most memorable. He had several ideas for flying, including a machine with bird-like wings which were spring operated. His man-powered ornithopter was, however, more typical of his theories, and to some extent his ideas were used – with the benefit of modern materials and technology – in the quest to build man-powered aircraft in the 1960s, when some of them were quite successful.

Real progress in ideas began to be made when men put aside the notion of simply emulating birds and flying themselves. The next ideas to evolve – which led ultimately to the development of the modern aircraft – were for devices which would be capable of getting up into the air and carrying men with them.

A Jesuit priest, Francesco de Lana, produced the earliest ideas for a man-carrying machine. His aerial chariot was illustrated and described in his own treatise on the subject, *Prodrono*, in 1670. De Lana's proposal was to have a boat-shaped craft suspended from four vacuum globes – spherical chambers of wafer-thin metal from which the air was extracted by vacuum pump. This would make the chambers lighter than air and so they would rise in the sky, taking the craft with them. Once

Leonardo da Vinci's fifteenth-century wing-flapping ornithopter design re-created as a full-size replica

airborne a square sail suspended from a central mast would give lateral movement. This remained entirely a theory, of course, for many impractical problems were obvious. There was no way of making vacuum chambers, either light enough or strong enough to avoid collapse, and no way of controlling height or direction. De Lana did not seriously expect his design to be tried out, but he was definitely the first to suggest the war potential of such an aerial chariot. He wrote '. . . who can fail to see that no city would be proof against surprise, as the ship could at any time be brought above its squares or even the courtyards of its dwellings, and come to earth so that the crew could land. In the case of ships that sail the sea, the aerial ship could be made to descend from the air to the level of their sails so that rigging could be cut. Or even without descending so low, iron weights could be hurled down to wreck the ships and kill the crews; or the ships could be set on fire by fireballs and bombs. Not only ships, but houses, fortresses and cities could be destroyed, with the certainty that the air ship would come to no harm as the missiles could be thrown from a great height.' Allowing for the much changed nature of twentieth century weapons, de Lana's predictions were remarkably accurate.

Not long after de Lana, in 1709, another Jesuit priest, Laurenco de Gusmao, took the idea a stage further when he built and demonstrated a model hot air balloon to the king of Portugal. This was a small replica of a proposed man-carrying balloon device, *Passerola*, which was clearly one of the first recorded examples of the hot-air balloon principle. Hot air rises, and in this type of balloon the envelope is open at the base with some type of burner arranged beneath the opening so that it heats the air in the envelope, causing it to lift. Passengers are transported in a basket beneath the envelope.

Passerola was to have a basket shaped like a giant bird, with head and tail and side 'wings' for stability. The full-size version appears not to have materialised, and it is recorded that the model version caught fire in mid air—it was being demonstrated inside the royal palace—and crashed to earth setting fire to some of the furnishings in the process. Hot-air balloons have come to grief like this many times since—one of the major hazards of this type of aerial transport.

The balloon goes up

It was in 1783 that the really significant step forward in the development of flight took place. This year saw the practical realisation of the hot-air balloon idea by the Montgolfier brothers, Joseph and Etienne, who were by profession paper makers from the town of Annonay, near Lyons. Conflicting accounts exist of what inspired the Montgolfiers. Some say they noted how ashes of burning waste paper were wafted up the chimney by the rising hot gases from the fire; others say that Joseph's shirt filled out and floated upwards when placed before the fire to dry. Suffice to say that they carried out some successful experiments indoors in November 1782. Small silk and paper bags held over the fire to fill with hot air were released and rose to the ceiling. This led the brothers to make a full-size

balloon, about 36 ft in diameter. On 4 June 1783, they carried
out the first outdoor test, filled the balloon over a specially-made
fire and let it soar away. It rose some 6,000 ft before the hot
air cooled, and it returned to earth about 1½ miles from the
take-off point.

The story spread far and wide and Etienne was asked to bring
a balloon to Paris to demonstrate before the king and queen.
The balloon used on this occasion was about 70 ft high and
had a wicker basket slung below it. By way of experiment, a

sheep, a cock, and a duck were placed in the basket and the balloon was released, descending to earth with the animals intact some 2 km distant. The date of this first animal-carrying flight was 19 September 1783. Better was to come, however, for the next stage was to be a manned flight. A 26-year-old volunteer, Jean-François Pilatre de Rozier, became the first man to go up in a balloon on 15 October 1783, when he rose to 85 ft in a tethered flight and kept it airborne for nearly five minutes by suitably tending the fire. Five more tethered flights at increasing heights up to 330 ft followed this, the last with a passenger on board. Then, on 21 November 1783 came the great day when the manned balloon made its first free flight. De Rozier, with the Marquise d'Arlandes as passenger, took 25 minutes

below, left
De Roziere's ill-fated hybrid balloon, using both hot air and hydrogen, was built for cross-Channel flight in 1785

above
The perils of ballooning: the Royal Nassau balloon of 1849 collides with a house

voisin. Sculp

on this historic flight, and they landed 5½ miles away from the starting point.

Meanwhile the French Academy of Sciences commissioned a physicist, J. A. C. Charles, to make an improved balloon. This led directly to the development of what we now know of as a hydrogen balloon. Hydrogen was a fairly recent discovery (in 1765) and was considerably lighter than air. In the Charles experiments a balloon envelope was made from rubberised silk, 13 ft in diameter, with an entry valve for the gas. Hydrogen was made from vitriolic acid poured over iron filings. There were problems piping it into the balloon and preventing leaks in the fabric panels, but the hydrogen balloon had the advantage of requiring just a single filling, while a major weakness of the hot-air balloon was the need for a constant heat source to keep the gases rising. Sometimes the heat source set light to the balloon envelope with disastrous results—indeed de Rozier, the pioneer balloon pilot, lost his life within two years in just such an accident.

The first Charles hydrogen balloon made its initial ascent—unmanned—on 27 August 1783, three weeks before Montgolfiers' first public demonstration. The Charles balloon actually covered some 15 miles from Paris, landing in the village of Gonesse much to the bewilderment and terror of the inhabitants, who had, of course, not the remotest idea of its import or significance and had never before seen or heard of such a mystifying shape from the heavens.

Charles now made a larger balloon intended to carry two people and this was 27½ ft in diameter. The refinements included a valve and a hose for filling the envelope, a net over the top of the envelope and guy ropes suspended from it to enable a wicker basket to be hung below. On 1 December 1783, Charles and a colleague Marie-Noel Robert made their first flight. This was completed after two hours in the air and Charles made the next ascent alone. With only one person now in the basket the balloon shot up to 9,000 ft very rapidly before Charles was able to let out some of the gas to cause the balloon to descend. This experience showed early on the importance of ballasting, and hydrogen balloons subsequently used a combination of sand bags and judicious venting off of the hydrogen to aid ascent and descent. The Charles balloon proved its worth and set a style in hydrogen balloons which is virtually unchanged in detail and layout up to the present time. Hot-air balloons, in contrast, have been developed more completely and modern hot-air balloons differ in some detail (though not in principle) from Montgolfiers' original.

After these pioneer hot-air and hydrogen balloon flights there was huge publicity and much further activity. Balloons in France were called originally 'Aerostats' when made on Montgolfiers' hot-air principle, or 'Charliere' when made on Charles' hydrogen principle. The hydrogen balloon actually eclipsed the hot-air balloon in a very short time, being more controllable and less prone to the risks from the fire. In fact it is only in recent years that the hot-air balloon has come back into its own for sporting purposes, being cheaper to make and maintain than a hydrogen balloon in modern economic conditions. In 1784 the first balloon flights were made in Italy and Scotland and in that year, also, there was a first flight by a woman at Lyons, France. On 15 September 1784, the first

First flight of Charles and Robert's Hydrogen balloon, 1 December 1783

JUVENILE FETE AND BALLOON RACE AT CREMORNE GARDENS.

balloon flight was made in England, by Vincent Luardi, piloting a hydrogen balloon from the artillery depot at Woolwich.

On 7 January 1785 came the most imaginative and spectacular balloon voyage yet – a crossing of the English Channel by Dr John Jeffries of America and J. P. Blanchard, a Frenchman. They took off from Dover, at the Shakespeare Cliff, and just managed to make the French coast, but only after having jettisoned all their ballast, personal equipment, and even some of their clothes in an attempt to keep airborne and clear of the ever-threatening sea. A notable feature of the flight was Blanchard's attempt to obtain a degree of directional control by fitting oars and a rudder to the basket, though clearly with only minimal effectiveness. In an attempt to improve on this performance, Montgolfiers' pioneer pilot, de Rozier, produced a hybrid balloon made from a hydrogen-filled envelope above a hot-air envelope. The crossing, from France to England, was arranged for June 1785, but the fire for the hot-air envelope ignited the hydrogen above and the entire craft crashed in flames before the French coast was reached. De Rozier and his companion were killed in what was probably the first recorded flying accident.

Balloons have always retained their place in the world of aviation and, indeed, there has been a mighty revival in recent years of the hot-air balloon. Most ballooning in recent decades has been of a recreational or sporting nature, though balloons have also been used for scientific experimental work.

It was not always thus, however, for in the nineteenth and early twentieth centuries the balloon had some importance as a weapon of war. Early use was made by the French of tethered hydrogen balloons for observation purposes during the Napoleonic wars. Several lithographs of the 1790s show them in service. A fanciful lithograph used as propaganda by the French showed Napoleon's planned invasion of England – with a huge fleet of balloons carrying soldiers over the English Channel, while more soldiers tunnelled underneath it and many more were transported by ship. In the American Civil War (1861–65) specialist balloon companies operated hydrogen balloons for reconnaissance purposes. They were genuinely useful, giving the field commander a pair of 'eyes' far more extensive than that of even the fleetest cavalry scouts. In the Franco-Prussian War there was a more novel use of the

above, left
The balloon craze in the nineteenth century: a balloon race for children at a fete in the Cremorne Gardens, London, in 1859

above
An early British balloon ascent. Capt. Paget of the Royal Navy and Mr Sadler rising above the Mermaid Tavern at Hackney August 1811

balloon–to beat the siege of Paris. Mail and important documents were flown out of Paris by hydrogen balloon and over the heads of the Prussian attackers. By the 1890s there were balloon units in most of the world's armies. In Britain the Royal Engineers attracted much attention by the novelty and smartness of their drill, and balloons were operated for observation purposes in the South African War of 1900. In the First World War there was balloon activity by both sides on the Western Front, and balloon crews were a brave breed of men, for the balloons became targets for marauding aircraft and were accordingly dispatched. The mode of operation was to have a field telephone cable attached to the tethering line so that an observation officer in the wicker basket could actually telephone such information as fall of artillery barrages and enemy movement behind the line.

By this time the balloon for military purposes had changed its shape–rather egg-like with inflatable fins and stabilisers incorporated in the 'thin' end. This gave the balloon much more stability when it was riding on a tether, for the stabilisers helped it head into the wind. This new idea was evolved in 1897 by a Captain Perseval of the Prussian army. The shape remains unchanged to this day. In the Second World War, the balloon had been rendered both obsolete and too vulnerable for observation use, but found a new lease of life as a barrage balloon, supporting cables and deterring low flying bombers. In recent years military balloons have mainly been used for meteorological and scientific work, though trainee parachutists get their first taste of 'jumping' from a basket carried under one of the self-same 'kite' balloons.

One other big change over the years has been the replacement of the very unstable hydrogen by coal gas (in some early cases) and by helium.

Towards powered flight

One of the greatest theorists in the quest for practical, powered flight – or indeed non-powered flight – was Sir George Cayley, one of Britain's many prolific scientists active in the nineteenth century. Born in 1773, Cayley enumerated or forecasted in his time most of the principles of powered flight which have since been developed, including the jet engine, the airship, the importance of dihedral and other aerodynamic factors in sustained flight, streamlining, stability and loading factors, internal-combustion engine power and the 'convertiplane' idea (vertical take-off and landing).

Cayley's first aircraft design was produced in 1799, although it was not actually built and flown, but by 1804 he had designed and flown a small glider. Cayley was quick to appreciate that flapping wings in the manner of a bird were not essential for sustained flight. What was required was lift and power and these two requirements could be separated so long as they were both available. Cayley based his ideas on the age-old kite, a favourite child's toy said to have been first invented thousands of years ago by the ancient Chinese.

Cayley's first aircraft design was essentially a kite attached to a pole, with a small tail in the form of sails arranged in a cross and pivoted. The movement of the tail could thus give directional guidance while the main 'kite' part of the machine – the wings – gave stability and lift. The wings were arranged at a shallow angle in a flat V configuration, according with Cayley's own postulation that dihedral (as this was called) aided stability in flight. The glider worked perfectly in flights down a hillside and Cayley went on to make a version twice the size, with equal success.

Before his death in 1857 Sir George Cayley produced even more significant designs. In 1842–43 he published plans for what was essentially a multi-rotor helicopter. This 'aerial carriage' had four contra-rotating, multi-bladed rotors, plus two further propellers at the back to give forward motion. The design was brilliant and practical except for one thing, the lack of a really powerful engine. Cayley proposed a vertical boiler steam engine, but lack of any suitable small unit of this type prevented any working prototype being made. It is of interest to note, also, that an Italian, Vittorio Sarti, had devised a helicopter type of design in 1825 in which jets of steam played on the rotors from a boiler underneath. This existed in model form but was not built so far as can be determined.

A contemporary of Cayley's, William Samuel Henson, produced in 1842 a patent design even more remarkable in its way, for it was extremely close in concept and even appearance to airliners of 80 or so years later. Known in its time as an 'aerial steam carriage', Henson's machine had a ribbed, fabric covered wing exactly like that subsequently used in later aircraft, and the wings were supported by king posts and trusses in precisely the manner commonly adopted years later. There was a wingspan of 150 ft and propeller rotors 20 ft in diameter, arranged in 'pusher' configuration behind the wings. Slung below the wings was a wood passenger cabin which also housed the steam power unit. The undercarriage, of tricycle type (again years before its time), was attached to the fuselage. More than any other nineteenth-century design, Henson's 'aerial steam

A Serbian Army observation balloon in 1908 manoeuvres

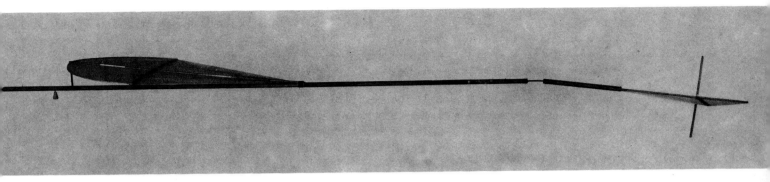

carriage' gave a prophetic glimpse of the future of air travel. Unfortunately, however, the realisation of the project did not match the promise. Henson was initially ambitious and in the manner of many another Victorian transport entrepreneur he promoted a grandly titled company with an equally grand prospectus; The Aerial Transit Company was the name, and it issued literature promising (and illustrating) flight from London to Paris and even on to Egypt. All this happened before the aircraft had actually flown.

Aided by a friend, John Stringfellow, Henson made a 20 ft wingspan model of his 'aerial steam carriage' in 1847 but its first test flight ended in failure—just a short hop at the end of the launching ramp before plunging to earth. The reason was the same as before, lack of power in the steam engine. Humiliated and disillusioned at his failure, Henson withdrew from the scene and made no more than this one memorable contribution to aviation history.

John Stringfellow, Henson's friend, was more persistent and more successful. Basing his design on Henson's layout, Stringfellow had a 20 ft wingspan, steam-powered model flying in 1848 but it was launched and supported from a wire line and was never really a 'free' flyer. A German army officer, Werner Siemens, patented a design at about the same time—a steam-driven monoplane with wings which flapped in ornithopter fashion. But there is no evidence to suggest it was actually built and flown.

Sir George Cayley returned to the aircraft scene in his declining years (after some time concentrating on airships) and he achieved spectacular results with two gliders. In 1849 he completed a triplane glider with flapping wings, the moving wings being intended to give a vestige of power assistance in the glide. It certainly worked and a replica of it was built in recent years by an airline company. Such flights as the triplane made were piloted by boys—short hops down a hillside. Much more ambitious was a splendid 'new flyer', built and flown in 1853. It had a bath-shaped gondola, a tail to give directional steering, a wheeled undercarriage, and two flapping wings worked by hand levers from the gondola. Sir George's coachman was detailed off as pilot and the glider was pushed down a hillside and became airborne in a short hop across the valley—the coachman is believed to have been the first man ever to fly in a heavier-than-air machine. That this was no mere fluke or fable was proved in 1974 when a British TV company built this glider all over again from Cayley's original plans and descriptions and it flew before the television cameras. The replica is now in the Science Museum, London, a fine tribute to a true pioneer of the aviation world.

A modern replica of Cayley's first glider of 1804

opposite
A balloon fantasy: this was the aerial exploration balloon 'Minerve' proposed by Dr Robertson in 1803 and much caricatured in the popular press

In 1857 two Frenchmen took some honours. Jean-Marie le Bris built a remarkably streamlined, needle-nosed glider which was launched from a horse-drawn cart towed at speed. The aircraft lifted off briefly. Felix du Temple, a French naval officer, patented an advanced monoplane design with front engine, tricycle undercarriage and aluminium cladding. Du Temple made a model which flew successfully with both clockwork power and a steam engine. In 1871 a full-size version of this machine was built with a hot-air engine. Du Temple had one of his sailors pilot the machine, launching it down a ramp so that it did make a short 'hop'.

In 1869 Stringfellow again entered the aviation world, this time with a triplane which was very similar in configuration to a box kite. Power to two 'pusher' propellers was by belt drive from a centrally-mounted, light steam engine. Like his earlier machine, however, this new one was also flown only on a support line. It foreshadowed the 'box-kite' form of wing configuration much favoured by the next generation of aviation pioneers.

Over the remaining decades of the nineteenth century, pioneer aviators became surprisingly numerous, with men from many lands trying their ideas to develop practical aircraft. In all cases, however, they were thwarted by the lack of a suitable engine of reasonable power-to-weight ratio. The steam engine was, with few exceptions, the only available choice. The actual aerodynamics was proving less of a problem. Clearly, had the internal combustion engine been available sooner, such machines as Henson's in the 1840s might have been a practical proposition and the whole history of aviation might have been very different. Fertile ideas abounded. In 1867 two designers, Butler and Edwards, proposed a delta wing, rocket-powered aircraft, with the rocket carried in a nacelle, which could be slid back and forth beneath the delta wing to vary the centre of gravity and thus give a degree of longitudinal control. In 1875 a Frenchman, Tatin, designed a very clean looking monoplane with two propellers driven by compressed air. This overcame the power-to-weight problem of the steam engine, but

below
The failure of Henson's design was ridiculed by contemporary cartoons

below, right
Lilienthal's glider of 1896

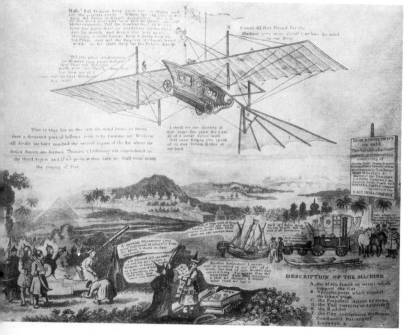

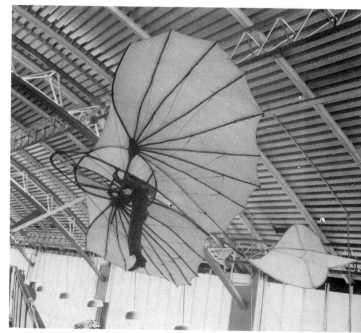

Hiram Maxim's steam flying machine of 1894 on its launching track

compressed air was not really a practical proposition and the project did not mature. The same year Thomas May, a founder member of the Royal Aeronautical Society, successfully flew a small, twin-engine, steam monoplane tethered on a circular track. The maximum height achieved, however, was no more than 6 inches.

Clement Ader, in France, had a significant (but not exceptional) success with two steam-powered aircraft of bat-like appearance. *Eole*, the name of the first one, actually flew some 150 ft on 8 October 1890, but the aircraft had no form of directional control. A second aircraft, *Avion III*, had two engines but did not fly so well. Ader had a claim to be the very first man to fly in a powered aircraft, though as he had no means of controlling it, his claim is not generally considered by historians.

Sir Hiram Maxim, inventor of the Maxim gun, was one of the most successful of aviation pioneers. Realising the limitations of steam power, he built a machine with lifting surfaces commensurate with the $3\frac{1}{2}$ ton weight of the aircraft. The wing span was 104 ft, the length 110 ft. Special lightweight steam engines, three in number, were fitted in the nose. There was a skid undercarriage, twin rear propellers, and a triple arrangement of the wings. An elevator was carried on an extension forward, and a stabiliser aft. For tests the huge aircraft was run on a trackway, but it developed so much power that it lifted and broke away from the track after a 200 yard run, at which time its speed was 40 mph.

In the last ten years of the nineteenth century much successful experimentation was done with gliding. Otto Lilienthal was the most successful exponent. He perfected a hang type of glider, built several models and made over 2,000 flights before his death (in a crash) in 1896. His type of glider was essentially similar in principle to the modern hang glider which is, in fact, largely inspired by Lilienthal's work. In the USA Octave Chanute also produced successful hang gliders, and an Englishman, Percy Pilcher, was similarly active in this field. One of Chanute's assistants, Augustin Herring, left Chanute and produced a design for a compressed air driven version of one of Chanute's type of hang glider. It did not in fact fly but the configuration in box-kite form was later followed by several other pioneer designers.

21

Pilcher's glider 'Hawk' in which he died in 1899

Another important personality in verifying the excellent lift characteristics of the cambered surfaces of the box-kite configuration was an Australian, Lawrence Hargrave, who had flown and publicised large box kites in 1894. A contemporary of his in England, Horatio Phillips, developed a cambered wing and proved its lift characteristics with some experimental multiplane designs. In these the wings were narrow cambered units arranged rather like the slats of a venetian blind. Both Hargraves and Phillips (who patented the 'Phillips Entry') had independently shown the crucial necessity for a suitable airfoil shape to provide the lift characteristics of a moving wing shape.

Of the glider pioneers, Percy Pilcher was arguably the most inventive. He improved over the efforts of his contemporaries by fitting a light, wheeled undercarriage to his 'Hawk' glider of 1896 and by the following year, 1897, he was making quite long flights (up to 750 ft) using teams of men or horses to tow him from the ground—the tow line gave sufficient ground speed to make 'Hawk' airborne. By 1899 Pilcher had taken his ideas a step further forward. Aware of the limitations of steam power he took advantage of the recent introduction of compact internal combustion engines to design an oil engine of his own. It was of only 4 hp and drove a 4 ft diameter propeller. The idea was to fit it to one of his gliders to produce a practical powered aircraft. It might well have worked except that Pilcher himself was killed at the end of September 1899 when 'Hawk' crashed during a demonstration flight. With his death this line of development came to a premature end and it was to be several years before the dream of powered flight was finally realised in the United States.

From balloon to airship

If aircraft were—literally—'slow' to get off the ground in the nineteenth century, the development of the airship, from the original balloon concept, was surprisingly straightforward.

As early as 1816 Sir George Cayley produced a design for an airship, which was in essence an elongated balloon from which was suspended a boat-shaped gondola. In this there was a steam engine which operated flapping wings set in two banks, each side on outriggers. This design was never realised, but it pointed the way quite prophetically to the future.

In 1837, Cayley produced a more advanced idea for a steam-powered airship featuring propellers which could be moved to give steering. This was not built, but a Frenchman, Pierre Jullien, constructed a working model in 1850 which followed Cayley's principle quite closely, though the motive power was a clockwork motor. This was called 'Precurseur', a suitably prophetic name. It led another Frenchman to build a full-size airship which actually flew.

This was Henri Giffard's non-rigid craft which flew on 24 September 1852, in Paris. Giffard's craft was essentially an elongated hydrogen balloon with a gondola slung beneath in which was a small 3 hp steam engine driving an 11 ft diameter propeller. A canvas triangular sail gave directional stability. A grapnel was carried allowing the aviator to haul himself down provided the grapnel could catch tree tops or bushes. The initial flight covered 17 miles, at some 6 mph.

There was no real steering control in Giffard's airship, and this problem was not overcome until 1884 when two Frenchmen, Arthur Krebs and Charles Renard, built an electrically-driven airship 'France'. A special, lightweight, 8 hp electric motor with light batteries was fitted to the gondola in an attempt to gain a weight advantage over steam engines. As it turned out, however, the electric motor was heavier than Giffard's little steam motor. A model airship, also powered electrically, had previously been demonstrated (in 1881) by two French balloon experts, the Tissandier brothers. 'France' was reasonably successful with the great advantage that it could be steered and controlled without regard to wind direction. Lack of a suitable type of engine was the one factor holding up airship development, just as it held up the development of powered aircraft.

The first airship with an internal combustion engine was built in 1888, almost as soon as the Daimler motor first became available. This airship had a tiny, crude, 2 hp Daimler engine and was built by Dr Karl Wolfert.

Early attempts to build efficient, navigable airships were now dogged mainly by minor technical problems. A German, Schwarz, built an airship in 1897 which crashed disastrously. But in the same year a wealthy young Brazilian, Santos-Dumont, visited Paris and was bitten by the ballooning pastime after witnessing a demonstration. He ordered two balloons at once and named the smaller one 'Brazil', and the larger one 'Vaugirard'. The latter could carry five people, but the former was very much a personal balloon able to lift only 180 lb and folding up to fit into a hand valise.

Santos-Dumont was also an active and enthusiastic motor-cyclist—actually, he owned a motor tricycle. One day he realised that by combining the tricycle engine with a balloon he could have his own personal airship. After studying the subject he resolved to build a dirigible (rigid-framed) airship of elongated shape able to lift himself and the engine. In a hired mechanic's workshop in Paris the engine was adapted—$3\frac{1}{2}$ hp with a weight of 66 lb. The envelope of the airship was made of Japanese silk, pointed in shape at each end, $82\frac{1}{2}$ ft long, $11\frac{1}{2}$ ft in diameter. The gondola was suspended on wood struts which were attached to the rigid frame supporting the envelope. A rudder consisting of a steel frame covered with silk was attached beneath the

below, left
Hargrave's box kite of 1894–95

below
The gondola of the electrically-powered airship 'France' built by the Tissandier brothers

tail of the craft. Two bags of ballast were hung from the envelope, one forward and one aft. These were moved back and forth to shift the centre of gravity and so cause the nose to point up or down as required. The engine drove a propeller fitted at the back end of the cigar-shaped gondola.

Named 'Santos-Dumont No.1', the craft had its trial flight at Paris Zoo on 18 September 1898. Because of inexperience on the part of Santos-Dumont, the first journey was abortive. The craft was blown against some nearby trees and the envelope was ripped. In two days, however, the damage was repaired and the first flight took place. Santos-Dumont manoeuvred the ship in all directions with complete success. He then took it up to 1,300 ft. When descending, however, the gas in the envelope contracted so rapidly that the air pumps fitted for emergency inflation were insufficient to compensate for the loss of hydrogen. The envelope began to empty more rapidly and the airship plunged earthwards out of control. Santos-Dumont saw some boys playing below, threw them his groundline, and they pulled him safely to the ground.

'Santos-Dumont No.2' was an altogether larger airship, though similar to No.1 in construction and design. A small ventilator or rotary fan operated by the motor now supplemented the original pneumatic air pump. An 'interior' air balloon was now sewn inside the envelope and the fan injected air into this. The trial was fixed for 11 May 1899, again from Paris Zoo. While inflation was taking place it began to rain. Dumont decided to press on regardless and ascended. As he did so the cold, wet state of the envelope caused the gas to contract and the airship descended fast. Before Santos-Dumont could switch on the emergency pump, the craft struck some trees and was wrecked.

Undeterred and obstinate, Santos-Dumont pressed on. His airships Nos.3, 4, 5, and 6 followed in rapid succession. In the hope of avoiding future hydrogen contraction problems, he switched to coal gas for No.3 and had a shorter, fatter envelope. Dumont was still not fully satisfied. In No.4 he fitted a new four-cylinder motor (the old motor tricycle engine had two cylinders) and lengthened the envelope. A Mr Deutsch of the Paris Aero Club had meanwhile offered a £4,000 prize to the first man to fly by air around the Eiffel Tower and return to the starting point at the club headquarters, St Cloud. Santos-Dumont resolved to try for this. His first attempt ended in failure when the motor cut out and the craft descended into a chestnut tree in a Paris garden. On 8 August 1901, a second attempt also ended unsuccessfully—but spectacularly—when there was a gas leak and the craft landed involuntarily on the roof of the Trocadero Hotel with the gondola containing Santos-Dumont dangling precariously down the side of the building. He was rescued by ropes dropped down from the roof. At once he ordered a new craft (No.6) and in three weeks it was built. On 19 October 1901, he actually achieved the objective, arriving at St Cloud with but 30 seconds of the half hour to spare. Subsequently, Airship No.6 was lost in a flight to Monaco. Santos-Dumont made seven or eight more airships in later years but they represented no great advance on No.6. Though very much an airship pioneer, Santos-Dumont was never more than a rich and enthusiastic amateur and he played no further part in airship development.

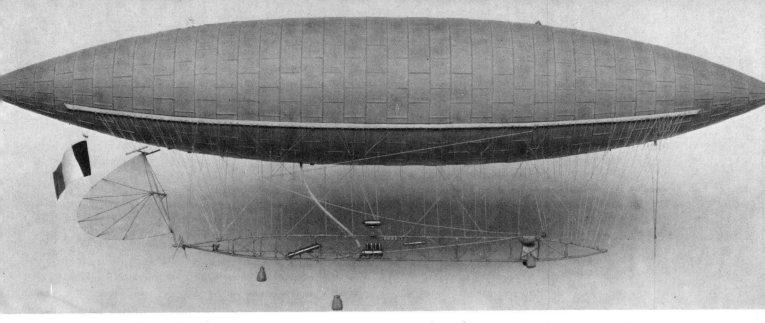

A model of airship pioneer
Alberto Santos-Dumont's '10' of
1901, closely similar in layout
to the No. 6 in which he made
his famous flight round the Eiffel
Tower in 1901

Meanwhile, in Germany, there was a contemporary figure at work, Count Ferdinand von Zeppelin, a retired army officer. Zeppelin was moved by the failure of Schwarz's airship in 1897 and resolved to perfect the idea. His previous experience had been limited to work as an observation officer in a hydrogen balloon during the American Civil War where he had served in the Confederate Army. Said Zeppelin after Schwarz's death, 'I intend to build a vessel which will be able to travel to places that cannot be approached – or only with great difficulty – by other means of transport, to undiscovered coasts or interiors, in a straight line across water where ships are to be sought for; from one fleet station or army to another, carrying persons or despatches; for observation of hostile fleets or armies, not for actual participation in actual warfare. My balloon must be able to travel several days without renewing provisions, gas, or fuel. It must travel quickly enough to reach a certain goal in a given number of days and must possess sufficient rigidity and non-inflammability to ascend, travel and descend under ordinary conditions.'

This statement was generally received at the time with some scepticism, even ridicule, for it was thought that an army officer could have no real idea of what could be achieved in the air. The project as conceived by Zeppelin was enormous for its day. To raise finance, Zeppelin sold his family estate and realised all his other assets to provide a total of £30,000. A company was formed under the patronage of the Kaiser and £40,000 capital was eventually raised. The new airship was to be bigger than any previously made – it had to be to fulfil Zeppelin's promise of performance – and, to facilitate handling, a huge floating workshop was constructed on Lake Constance. A timber shed buoyed on 95 pontoons was floated on to the lake and the airship was actually built in the shed over two years starting in 1898.

When finished, the craft was 390 ft long, 30 ft in diameter and had an envelope of rubberised silk. It covered an aluminium framework consisting of 17 polygonal rings at 24 ft spacings, linked by longitudinal aluminium stringers. There were 16 smaller compartments inside the framing, each containing a spherical, hydrogen-filled balloon. The free space in each compartment allowed air to circulate freely and provided room for the gas to expand when the craft operated in high temperatures. The other big advantage of this arrangement was that if one balloon deflated the loss of gas was kept minimal

British naval rigid airship R.23 of 1918 making an experimental air launch of a Sopwith Camel

A Zeppelin of the First World War, the L12, shot down and captured in the North Sea in August 1915

and stability could be maintained. The 16 balloons together had a capacity of 324,000 cu ft of gas. It cost about £500 merely to inflate the balloons. Two gondola cars, each 5 ft wide and 3 ft deep, were slung beneath the envelope. Each car had a 16 hp Daimler motor which operated two propellers carried above the car. Benzene fuel for 10 hours flying could be carried aboard. A steel keel joined the two cars and a sliding weight, hauled back and forth by cables, moved the centre of gravity to point the nose up or down. This obviated the use of ballast or the discharge of gas to adjust height. By 2 July 1900, the first trial flight took place after twice being postponed. There was trouble with the motors, the sliding weight, and many other functions. The airship was flown some three miles out, then returned, and was slightly damaged when coming up to her floating hanger. A second trial flight later in the month was more successful.

This first airship (by now known as a 'Zeppelin') was subsequently scrapped and a bigger and better one was started. But by now the money had run out and Zeppelin was forced to abandon his work. Private friends provided funds at the last minute and the new ship was finished in the summer of 1903. It had many improvements, including new, more powerful engines. The envelope was 414 ft long and the 16 internal hydrogen balloons (or bags) had a total capacity of 368,000 cu ft.

Mishaps once again attended the trial flight on 30 November 1903. It was towed from the shed by motor boat but a good following wind caused it to overtake the boat, which cut the tow rope. The weight of the rope pulled the nose of the craft down into the water. Then there was a collision with a pontoon and the net result was a six week 'docking' for repairs.

On 17 June 1904, it was once again airborne, when a height

of 1,500 ft was attained but moored overnight it was severely damaged in a gale and had to be broken up.

However, further Zeppelins were built, each more powerfully engined than before. They demonstrated the reliability and practicability of von Zeppelin's original concept and, at the time, Germany held a commanding lead in air power of this sort. By 1914 provincial airship travel within Germany was well established.

Zeppelin, having produced passenger carriers, went on to build purely military airships. Naval airship L1 was a milestone design built in 1911. L1 carried 17 hydrogen bags within her framework. The ship was 514 ft overall, 47 ft in diameter and of 776,000 cu ft capacity. Three Maybach engines gave 540 hp in total and a speed of 50 mph could be attained under good weather conditions. The old sliding weight arrangement was replaced in L1 by two pairs of plane-sets–like the hydroplanes of submarines. These were attached to the polygonal rings of the framework. The two gondola cars were joined by a catwalk and the keel was enclosed in rubberised cloth. The forward car contained the controls and navigation instruments, and two engines. The after car housed the third engine. There were gun positions atop the airship and also in the gondolas.

The Zeppelin design set the pace and rigid-type airships became the most favoured for new construction. The non-rigid airship as built by Santos-Dumont, however, remained popular and smaller craft were often of non-rigid type. In later years (including the Second World War and after) non-rigids became the only type of airship to be built.

The airship story developed very little beyond Zeppelin's masterly concept. Rigid airships got bigger and bigger, and Zeppelins fulfilled many of their designer's prophecies during the First World War. Night bombing of Britain became a dread reality, until British fighter planes got the measure of the tactics needed to meet these giant attackers on dark nights. After 1918 there was something of an airship race between the major powers, Germany, USA, and Britain, to build the most prestigious types. R-100, R-101, 'Hindenburg', 'Graf Von Zeppelin', 'Macon', 'Akron', and others were among the best known. But a series of spectacular and tragic accidents, notably the awful conflagrations of 'R-101' and 'Hindenburg', brought

US Navy L-type dirigibles in formation over California in 1944

a stop to all commercial airship development and operations in the late 1930s.

In the Second World War, small, non-rigid airships, mainly built by Goodyear, were extensively used by the US Navy for anti-submarine and patrol work, tasks at which they proved very suitable. In post-war years even most of those disappeared. The 'blimp' lingered on in post-war years, a last link with pioneers like Santos-Dumont and Zeppelin (indeed, Ballon fabrik Augsburg, which made some new envelopes for Goodyear airships in the late 1950s, then claimed to be the oldest aeronautical company in the world). Apart from a few attempts by lighter-than-air enthusiasts, 'blimps' survive for use as advertising vehicles, suitably emblazoned, or as aerial TV camera platforms. Westdeutsche Luftbeerung built a few small ships, while Goodyear 'blimps' became familiar in many countries as they piled up trouble-free publicity flight hours in the 1970s.

The dawn of powered flight

By the start of the twentieth century, all the essentials for successful powered flight had been discovered. Aerodynamic theories were known, lift and thrust could be achieved, and directional control maintained. Last of all there was the four-stroke gasoline engine, vastly superior to the various steam, electric and compressed-air engines which had been available to the earlier pioneers.

The United States was the most fertile area for development and there were two major protagonists working towards efficient, powered flight, the Wright brothers and Samuel Pierpoint Langley. Langley was already a famous scientist when he became interested in the problem of powered flight. Among other accomplishments, he was secretary of the Smithsonian Institute and therefore well placed to keep abreast of the lastest scientific developments. He built a very successful flying model in 1896 (actually his fifth such) and it performed well. With a 14 ft wing span, it was a monoplane with tail fin and what were, in fact, a second pair of wings at the tail. There was a 2 hp steam engine belt driving two 'pusher' propellers which were outigged from the fuselage between the main and aft pairs of wings. The model had all the recognisable characteristics of an airworthy machine in later days and was well ahead of its time. It was launched from a catapult and made flights over 400 ft long. Top speed was around 25 mph. Langley called his model the 'Aerodrome'.

In 1898 came the American-Spanish war, and the US Government, in the vague hope of finding some military use for the 'Aerodrome', funded Langley $50,000 to build a full-size version of his aircraft. In the actual 'Aerodrome', Langley took the important and fundamental step of replacing the steam engine with a lightweight, 50 hp gasoline engine. The 'Aerodrome' was completed in 1903, and on 7 October came the first flight. For reasons of safety, Langley chose a catapult launch from the roof of a houseboat in the Potomac River. Charles Manley was the pilot and he had designed the engine used. In the event the operation was a failure. During the launch the aircraft hit a post on the take-off ramp and plunged into the river. After recovering the aircraft, Langley rebuilt it and in December attempted a further flight; this time the back wings hit a post alongside the launcher catapult and the aircraft plunged ignominiously into the river again. By this time it was overtaken by further developments by other pioneers and no further tests were made. However, it is of interest to note that in 1914 the aircraft was actually rebuilt, re-engined, and successfully flown by another pioneer, Glenn Curtiss.

It was Langley's rivals, the Wright brothers, who finally took the honours for the first powered flight just a week or so after the second unsuccessful launching attempt of Langley's 'Aerodrome'. The memorable date of the Wrights' first powered flight–the start of a new era in transportation–was 17 December 1903. Undisputably, the Wrights were the pioneers of flying in its modern, recognisable form. While admired for their astuteness and perseverance in producing a practical aircraft, the major secret of their success was the ingenious control system which gave them complete directional command of the machine. Most previous aircraft of any sort–gliders, balloons, airships–depended on shifting ballast or a rudimentary rudder to give a change of direction. In the case of hang gliders of the type built by Lillenthial or Chanute, for instance, directional control was at its most primitive, depending entirely on the intrepid pilot leaning in the appropriate direction or throwing his legs sideways or backwards in the manner of a long jumper.

The Wrights had a better idea. They had been working on it since 1896, when, as cycle makers in Dayton, Ohio, they were first attracted to the problem of flight, whose early pioneers were receiving more and more publicity in the popular press. By 1899 the Wrights had perfected and flown a 5 ft wingspan glider. They perceived that a way of controlling the direction of flight was by warping the wing tips while the craft was moving through the air. They did this remotely on the kite by control wires worked from the ground, rather in the way that a modern, control line model aircraft–or, indeed, an elaborate box kite–is still operated today. The idea was successful, so they built a much bigger kite (or glider) of some 17 ft wingspan. This they shipped to the Kill Devil Hills, at Kitty Hawk, North Carolina, where they leased a work shed and had sand dunes all around them to ease the damage to craft and man in the case of any accidents. The first glider was flown in kite fashion and made a few flights with one of the brothers aboard. This was in 1900. In 1902, and again in 1903, they made many flights at Kitty Hawk in their second and third gliders.

Their research was thorough. They made a primitive air tunnel

Langley's full-size Aerodrome ready for launching on the Potomac River, 7 October 1903

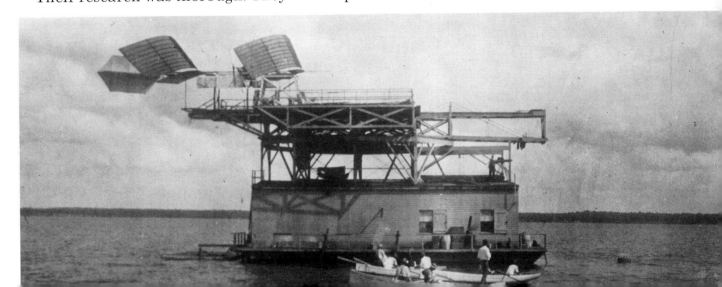

above
Horatio Phillips' Multiplane design. This 1904 version, which failed to fly, was one of a series produced by this English designer

top
The Wright brothers' historic first flight on 17 December 1903

at home and 'flew' models in it to determine the answers to the aerodynamic factors they needed to overcome for long duration flights. The warping wires for the wing tips were operated very simply by the pilot of the craft by the time the third glider was flying. The pilot lay prone facing forward, his hips in a dish-shaped rest on sideways slides. The warping control wires were led to this rest. Moving the hips to left or right moved the dish rather in the manner of a sliding seat in a racing skiff. This caused the control wires to shorten as appropriate to cause the craft to turn to left or right. The wind tunnel tests showed how the aircraft could be made to ascend or descend. This was done by placing an elevator (like a small wing) on outriggers forward of the wings. Wires could cause it to tilt back and forth. A vertical rudder was added on outriggers behind the wings to assist in turns and stop the craft tiltling too far each side on the turn.

With their third glider performing well, after hundreds of flights, the Wrights felt confident of achieving powered flight. They built an even bigger version of their craft called the 'Flyer' with a 40 ft 4 in wingspan. A suitable, lightweight, gasoline engine proved a problem, however, for there was nothing available to meet requirements. So they had one built. Then mechanic, Charles Taylor, carried out this work and built a 12 hp unit weighing about 170 lb. This was powerful enough to get the aircraft and pilot and engine airborne – an all-up weight of around 750 pounds. The engine was mounted on the lower wing, driving two pusher propellers by bicycle chains. The wind tunnel tests had taught the Wrights that even the propellers could help achieve lift by themselves being of aerodynamic cross-section.

On the morning of 17 December 1903, after a number of earlier mechanical set-backs (such as twisted propeller shafts), the Wright 'Flyer', with Orville at the controls, trundled along its trackway on a skid undercarriage and made its first bumpy 12 second hop into the air. The brothers were jubilant and made three more flights that day before a gust of wind turned the machine over and damaged it. The last flight of the day lasted 59 seconds and the 'Flyer' covered 852 ft. By the next year, the brothers were making flights round a circular course, and they went on to build a 'Flyer' No.2 and in 1905 they built No.3. By now they were flying from a field near their home town in Ohio and were achieving flights up to 24 miles long, around half an hour of flying time. They offered the services of their aircraft to the US Army, who declined with thanks, not believing the claims for the aircraft's performance. At about this time the Wrights lost interest in flying and the aircraft were stored in the belief that the potential for future development was limited.

The Fledgelings

Storing their aircraft was a short-sighted move on the part of the Wright brothers and left room for others to make progress. It was the Brazilian airship pioneer Santos-Dumont who took the honours to Europe. As enthusiastic as ever, he had Gabriel Voisin build him an aircraft which was based in part on Hargrave's box-kite configuration. Known as the '14 bis', the Santos-Dumont machine was one of the strangest yet. It appeared, even to eyes unaccustomed to aircraft, to be flying backwards. The wings in box-kite form had a fuselage linking them to large elevators and rudders in a forward structure. There was a 50 hp engine at the rear driving a pusher airscrew. The pilot—Santos-Dumont was a small man—stood between the wings with the controls to hand. In some respects, '14 bis' was an advance over the Wright design. It had a wheeled undercarriage which freed it from the restricting trackways and catapults employed for the Wright 'Flyer'. It had a more direct drive airscrew and the controls were less complicated. However, it was primitive and clumsy in construction with none of the light elegance of Wrights' design. First flight of '14 bis' was on 23

The classic Blériot XI, the definitive Blériot design

October 1906, when its first 200 ft flight won Santos-Dumont a cash prize for the first man to fly more than 25 metres in Europe (this was itself a second attempt, the machine failing to get airborne a month earlier). On 12 November 1906, Santos-Dumont made a more successful flight of 722 ft at a speed of over 25 mph. The big publicity accorded to Santos-Dumont's flights made Europe, and in particular France and Britain, into a hive of aviation activity. While the Wrights had been fairly quiet and self-effacing men, Santos-Dumont was an enthusiast who made his flight trials into well-advertised events.

The flights gave a big boost to the makers of '14 bis', the Voisin brothers Gabriel and Charles. They had been inspired by the efforts of Clement Ader and had become professional builders of airships, gliders, and aircraft, with Santos-Dumont as a good customer. The Voisins met Ader in the 1890s. They made it their business to study flying characteristics and their work in France paralleled to some extent the Wrights' work in the 1900–05 period. Gabriel Voisin produced some successful gliders comparable in size to the Wrights' but using the 'box-kite' principles for the wings. The most spectacular success was the first ever take-off from water when a motor boat towed Gabriel airborne from the surface of the Seine on 6 June 1905.

Learning from the crudeness of '14 bis', Santos-Dumont

designed a petite, lightweight machine in 1907 which he envisaged as an aircraft which could be sold to other pilots for home construction. Known as the 'Demoiselle', this fragile looking monoplane, with a bamboo and fabric structure, lays claim to being the first 'production' aircraft and the first 'private' light plane. Santos-Dumont saw it as an aircraft for everyman, amazing optimism and foresight for 1907! The aircraft was way ahead of its time. It had a high wing with the pilot sitting low and well forward between the wheels. A 25 hp Darracq car engine gave it a speed of around 60 mph. The only drawback was that the machine was built to suit Dumont and he was a small man weighing no more than 115 lb. So only small pilots could get the Demoiselle airborne. The point was well made in 1964 when a replica Demoiselle was made for the film *Those Magnificent Men In Their Flying Machines*. Built to the original plans and even with a more powerful engine, the aircraft had to be flown by a petite lady test pilot to get it airborne for the flying sequences.

The French were now making all the running in aviation development. A clutch of successful constructors and pilots made their name in the years following the Santos-Dumont flights. The Voisins came upon the scene with a really practical production design, still on the box-kite principle, but no longer a 'canard' with rear engine and forward elevators as in '14 bis'. The 1907 Voisin aircraft 'Henry Farman No.1', which they built for Henry Farman, was a box-kite machine with engine and pilot between the mainplanes, a pusher airscrew, and outriggers supporting a box-kite tail at the rear and an elevator forward of the mainplanes.

This Farman Voisin design became one of the most widely used and successful aircraft of its day. Wing warping was used to control direction and the elevator for vertical control, as in the Wright 'Flyer'. A steering wheel on a tilting 'joystick' had become the almost universal means of control by this time, replacing the Wrights' ideas of levers and slides.

On 13 January 1908, Farman used one of these Voisin machines to win the Deutsch Archdeacon prise (50,000 francs) which had been put up for the first man in Europe to fly a closed one kilometre circuit. This was achieved at Issy-le-Moulineaux, Paris, which was the first purpose-built aerodrome in the world. Farman—an Englishman who had lived most of his life in France—had obtained permission from the military authorities to use some old parade grounds there as a centre for his activities. By 1909, Henry Farman had developed his own design of aircraft based on experience with the Voisin biplane. He got rid of the vertical panels from the tail surfaces—Voisin's relic of box-kite development—and replaced them with hinged ailerons, one on each wing. This device superseded the wing warping and had first been postulated by the American inventor Alexander Graham Bell. A lightweight, four-wheel undercarriage and wood skids was substituted for the heavy 'pram wheel' undercarriage of the Voisin designs. The Farman aircraft at first had a light car engine but he soon replaced it with a type which was to become famous—the Gnome, rotary, air-cooled engine. The Gnome engine was one of the fundamental developments in the history of aviation, for it was the first with a really satisfactory power-to-weight ratio which gave an adequate reserve of output for sustained flight by heavier

Most famous of the early aircraft engines was the Gnome rotary, in which the crankshaft was bolted rigidly to the aircraft structure, while the cylinders rotated round it. This seven cylinder Gnome of 1908 rotated at 1,200 rpm, developed 50 hp and weighed 165 lb

aircraft. The rotary principle reversed the conventional arrangement of an internal combustion engine. Instead of the cylinders being fixed so that the piston cranks drove the crankshaft, the cylinders of a rotary engine moved round the crankshaft as the pistons themselves gave rotary movement to the crankshaft. The arrangement was crude, but it saved weight and the movement of the cylinders through the air served to cool them. There were complications but all were overcome. The fuel mixture was fed in through the crankcase on later (Monosoupape) rotaries, and was forced into the cylinders by centrifuge as they spun round, with the cylinder ports opening and closing in turn. Exhaust gases were emitted through the far end of each cylinder.

The Farman aircraft with what were now quite revolutionary features became the most popular and successful in Europe, Farman eclipsing Voisin as a constructor. Henry Farman's brother Maurice also made aircraft. His design was not unlike Henry's, except that the elevator was attached to extensions of the undercarriage skids. He also provided an enclosed fabric 'cockpit' for the pilot.

On 25 July 1909, yet another French pioneer made the headlines when Louis Blériot, a car accessory maker, became the first pilot to fly from France to England. Blériot had been associated with the Voisins in their early glider trials, but had gone on to investigate a monoplane (single wing) powered design. Developing designs of his own was not without hazards; Blériot crashed several times but survived. In his eleventh attempt – the Blériot XI – his monoplane idea achieved great success. The London *Daily Mail* had offered £1,000 for the first cross-Channel flight and Blériot determined to win it. He made the 22 mile Calais-Dover trip in about 38 minutes and ended in typically spectacular fashion by crumpling the undercarriage in a heavy landing on Dover cliff top. Blériot's success excited all Europe. A feature of his design – which was simple to facilitate – was the moveable wing-tip type of aileron, an alternative idea to that used by Farman.

By now many other aircraft designers were coming up with new ideas or new aircraft – not necessarily at the same time. Among them was a neat all-metal framed monoplane by Robert Esnault-Pelterie, the REP; this was an advanced design for 1908, with radial engine (rather than rotary – the cylinders were fixed,

bottom
A modern flying replica of the
Bristol Boxkite of 1910

below
Blériot's arrival at Dover after his
historic first cross-Channel flight

The first international air race meeting at Hendon in 1913. All the aircraft are French: winner Claude Graham-White's Maurice Farman is passing over a Blériot and two Morane Saulnier monoplanes

though arranged round the crankshaft), wingtip wheels, and joystick control. An early crash by Pelterie caused him to withdraw from aircraft development as a result of injury. Another famous type was the Antoinette, a monoplane which was both successful, popular, and graceful. It had a 50 hp, steam-cooled, in-line engine designed by Levassaveur. It was the first plane to fly at 500 ft in a gale force wind. A British pilot, Herbert Latham, flying an Antoinette, had been Blériot's nearest rival to cross the Channel but crashed into the sea while making the attempt a week or so before Blériot's successful flight. In England a major pioneer was Alliott Verdon-Roe whose AVRO machine, a triplane, with a 9 hp JAP motor-cycle engine enjoyed great success. De Havilland and Blackburn were other pioneer names in England, though imported Voisin and Farman machines were used by many of the British pioneer flyers.

The 1908–09 period was something of a watershed year in aviation. Powered aircraft were now a proven proposition, France was the world leader with 12 or more makers and in England, USA and Germany in particular a new generation of pioneers attracted to flight were coming up with new ideas – both in aircraft design and, perhaps more ominously, in the potential and use of the new-found facility of flight.

At Rheims in 1909 a great international air meeting set the seal on achievements to date. Some 38 different aircraft appeared in eight days of flying events and demonstrations. Henry Farman made the longest flight – 112 miles – and Blériot the highest speed – 48 mph. One of the most successful competitors, who won the 20 km and 30 km races, was from the New World – Glenn H. Curtiss from Hannondsport, NY, who was taking over where the Wrights left off. In 1908 Curtiss had already won a prize offered by *Scientific American* for the first, officially-observed, powered flight in USA. His home-built aircraft was 'June Bug', and we meet Curtiss again shortly involved in the next stage of aviation development – the world of military aviation.

War in the Air

The year 1908 saw several military authorities begin to take a new interest in aviation. Wilbur Wright took a Flyer to France and was asked to demonstrate it by the army there. Meanwhile, in America, the Wrights had received a US Army contract to deliver 'a heavier-than-air machine'. Orville Wright flew this Flyer at Fort Myer, in Virgina and on 9 September 1908 made the first flight of over one hour duration. On the same day he took up an army officer as a passenger—one of the first-ever passenger flights—and a week later, while making a similar flight, the aircraft crashed and the passenger, Lieut. Thomas Selfridge, died from his injuries, the first air crash victim. The next year the Wrights taught two army officers to fly, and on completion of the trial programme they sold the Flyer to the US Army for $25,000 with a $5,000 bonus because the machine exceeded the contracted speed on the trials.

Glenn H. Curtiss remained the Wrights' rival, and his aircraft, with wheeled undercarriage and a purpose-built, 50 hp engine, was technically more advanced than Wrights' Flyer design, which was still launched from a trackway. When the US Army bought another machine, it was a Curtiss. They carried out some early bombing trials with this, and then did some trials with live bombs from the Wright machine.

Also in 1910, the French Army purchased some aircraft and trained 60 pilots. One of their aircraft, a Nieuport monoplane, was the first to be armed with a gun. The first French military air exercises were held in 1911. That year, too, saw the first offensive use of aeroplanes in war when an Italian aircraft dropped bombs on the Turks in the war in Tripoli in November 1911. Aircraft were also used by the Italians for reconnaissance in this war.

Meanwhile, the period 1909–12 saw immense aviation activity in all the major nations as young men caught the flying craze, and meets, rallies, races, and demonstrations became great crowd pullers. Newspapers offered cash prizes for all sorts of achievement. In the London–Manchester air race of 1910, for instance, a pioneer British flyer, Claude Graham-White, became the first man to fly at night. While these activities were not directly connected with military affairs, they had an important function in allowing the pilots to explore the full potential of flight, and much that was learned then came into its own when military flying expanded at a colossal rate.

Britain's first connection with military flying came via an American, the flamboyant Colonel S. F. Cody. Observation balloons were used by the British Army from the 1890s. By 1906 Cody had become Kite Flying Instructor to the British Army by virtue of his ingenious large box kites which could support a man. A very large kite took up a cable and a man with a smaller kite was attached and was carried up the cable by the smaller kite. The idea was to use the system for observation. In 1907 Cody fitted an engine to one of the kites and got a grant from the British Army to build an aircraft. This was an unweildy machine with a Panhard engine. In May 1908 it made its first, short flight and on 16 October 1908, made its first 'official' flight at Laffan's Plain, Farnborough. To this day Farnborough has remained the 'home' of British aviation development as the Royal Aircraft Establishment.

left
A Wright military aircraft in flight at Fort Myer in 1908

below
Colonel S. F. Cody's flying machine of 1908

Sailors in the air

A few individuals began to suggest the potential value of aircraft in maritime operations, almost as soon as the Wright brothers and other pioneers made their first, practical, powered flights from 1903 onwards. It was not until September 1910, however, that any major attempt was made to place naval aviation on an official basis. Appropriately enough, it was the US Navy which made these first steps when the Navy Department assigned an officer, Captain W. I. Chambers, to the post of 'officer in charge of aviation', responsible for all aircraft affairs. (The British Navy had appointed an officer to take charge of air matters earlier than this, as outlined later, but at the time this was specifically in connection with 'lighter-than-air' machines – airships.)

Capt. Chambers had been one of a group of US Army and Navy officers who had watched official service demonstrations of a Wright flyer two years previously in 1908, and it was the favourable report of this demonstration which led to Chambers' appointment. The initial conception of the value of the aeroplane at sea was centred primarily on its 'scouting' ability – flying ahead of a fleet to overcome the centuries-old tactical limitation which restricted a naval commander's knowledge of the situation to his view of the immediate horizon.

The most successful aircraft then in production in America was Glenn H. Curtiss's famous 'Pusher' design which had a V8, 75 hp engine, a top speed of about 60 mph, weighed about half a ton, and was $28\frac{1}{2}$ feet long. Chambers enlisted the co-operation of Curtiss and arranged for one of his aircraft to be used in a demonstration flight from a US warship.

It was a civilian, Eugene B. Ely, a Curtiss demonstration and test pilot, who had the distinction of making the first ever take-off from a warship at sea. The demonstration arranged by Capt. Chambers took place in Hampton Roads, Virginia, mainly for the benefit of the Navy Board. A temporary wooden ramp was built on the forecastle deck of the 3,750 ton 3rd class cruiser USS *Birmingham*, the Curtiss Pusher was hoisted aboard and, on 14 November 1910, Ely made history by trundling the Curtiss-owned machine into the air from the anchored vessel, with a perilous dip seawards from the end of a 50 ft ramp.

A much more effective demonstration of the possibilities of aircraft operations from warships was quickly arranged by Captain Chambers, again with the aid of Curtiss and his test pilot Ely. A larger ship, the armoured cruiser USS *Pennsylvania*, was given a much wider, wooden landing platform, rectangular in shape, 130 ft long and 40 ft wide. This time it was erected over the after deck of the vessel stretching from the rear superstructure and overhanging the stern. The plan now was to land on as well as take off from the ship. *Pennsylvania* was anchored with other units of the fleet in San Francisco Bay on 18 January 1911, the day chosen for the demonstration, and Eugene Ely planned to fly aboard, then turn the Curtiss Pusher round and take off once more for the shore. At 10.45 a.m. on what proved to be a misty, dull morning, Ely took off from Selfridge Field just outside San Francisco and headed seawards towards the anchored ships, 13 miles away. Due to the poor visibility he had to keep low in order to spot the ships, and *Pennsylvania* sounded her siren continually as a signal to aid

Eugene Ely makes the first landing on a warship, the USS *Pennsylvania,* on 18 January 1911

Ely in picking her out from the other vessels. Having reached *Pennsylvania*, Ely flew at leisurely pace the whole length of the ship about 100 yards clear at deck level, then made a climbing turn to line himself up with the landing platform. Gently and with perfect judgment he dropped down over the stern at about 40 mph, cut his engine and plopped the Curtiss on to the wooden structure.

The problem of slowing up the Curtiss once it touched down had been simply but brilliantly tackled by Ely and Chambers. They fixed three hooks on to the main axle of the machine's undercarriage and stretched ropes, weighted at each end with sandbags, laterally over the rear end of the landing platform. To keep the ropes clear of the deck they used battens placed end to end in two fore and aft rows.

above
The Caudron G.III was used by the French and British air forces at the start of the First World War

top
A Fokker Eindecker manoeuvring to attack a French Voisin pusher plane

The hooks on the undercarriage would catch one or more of the ropes and the weight of the sandbags would bring the aircraft to a swift stop. The idea worked perfectly and the Curtiss was halted within 60 ft. Thus the basic principle used today for a carrier deck landing was demonstrated most effectively in the first shipboard landing of all.

Naval aviators, however, were to try many more complicated ways of deck landing in the years to come before reverting to a more refined method exactly akin to Ely's. The success of Ely's first attempt was celebrated by a chorus of sirens from the assembled ships. Immediately after this historic shipboard landing the intrepid Ely took lunch with the captain and officers of the *Pennsylvania*, turned his aircraft round, ran it to the full length of the landing platform and, just 60 minutes after the landing, took off once more for Selfridge Field. The aircraft dipped for a moment towards the water before picking up height, circling the ships and heading for shore. 'On arrival at Selfridge Field,' the British magazine *Flight* recorded, 'he was vociferously cheered by the officers of the 13th US Infantry Regiment who were in camp on the field.'

Eugene Ely, the cool and brilliant pioneer of shipboard aviation did not, alas, live long to bask in the glory of his great achievements. Nine months later he died in a flying accident at an air show in Georgia. He was just 25.

It is worth recording at this point that one of Ely's colleagues, John McCurdy, who was the Curtiss chief test pilot, was so inspired by Ely's first take-off, from USS *Birmingham* in November 1910 that he arranged to try a similar feat from the deck of the Hamburg-American liner SS *Pennsylvania* and had a similar type of wooden flying-off platform built on the stern of the ship. Though a Curtiss Pusher was loaded aboard for the flight towards the end of November 1910, the attempt was postponed and finally abandoned completely.

Despite these far-sighted experiments and effective demonstrations by Ely, the Navy Department was less impressed. The major objection to flying platforms was the 'inconvenience' from the tactical point of view, since erection of the wooden staging of necessity restricted the use of at least one of the ship's turrets. The Secretary of the Navy directed Captain Chambers to investigate the possibility of a naval aeroplane 'capable of landing on water alongside a warship and fitted for hoisting aboard'–in other words he preferred the use of seaplanes which did not interfere with a ship's fighting capability. At this time, however, there were no seaplanes–or 'hydro-aeroplanes' as they were known–in America. A

A Short S.41 seaplane of the Royal Navy in 1912

Frenchman had produced a practical seaplane in 1909–10 and this had inspired Glenn Curtiss (who had experimented unsuccessfully with a seaplane idea in 1908) to adapt the Curtiss Pusher as a 'hydro-aeroplane' by fitting floats in place of the nosewheel and under the wings. Curtiss carried out trials off North Island, San Diego, California, at the end of 1910 and soon found his initial flotation gear to be clumsy and crude–the under-wing floats were made from inflated motorcycle inner tubes. He therefore discarded floats and fitted a single large pontoon under the existing undercarriage, retaining the wheels. With a Pusher so modified, Curtiss made a first successful test flight on 26 January 1911, just a week after Ely's deck landing demonstration. Chambers sent two young naval officers to watch progress with the trials and at the same time to start learning the rudiments of flying.

On 17 February 1911, Curtiss gave an official demonstration to the Navy of the 'hydro-aeroplane's' capabilities. He flew out to the USS *Pennsylvania* in San Diego Bay, the aircraft was hoisted aboard by crane, then later lowered back into the water for the return flight to North Island.

This demonstration exactly fulfilled Navy Department requirements and the Bureau of Navigation (the Navy Department then responsible for air matters) was allocated $25,000 for the 1911–12 fiscal year specifically to spend on three aircraft. One was to be a standard Curtiss Pusher (as had been used by Ely) as a training machine, one was to be the modified 'hydro-aeroplane' version of the type Curtiss had demonstrated, and the third was to be a Wright Flyer modified by the addition of floats. The two Navy Officers, Lieut. Ellyson and Ensign Pousland, were to be taught to fly these machines by Curtiss.

As finally modified for the US Navy, the Curtiss Pusher 'hydro-aeroplane' had a retractable undercarriage which could be hinged up above pontoon level for water landings, or lowered to below pontoon level for ground landings–a true amphibian in fact. It was called the Triad because it could operate in three ways, in the air, on the ground, or on water. The US Navy gave it the designation A-1 (the A standing for aircraft). Completed at the Curtiss workshop and base at Hammondsport, New York, the A-1 Triad, the US Navy's first aircraft, was officially handed over on 1 July 1911. The next day, Lieut. Ellyson received Navy Aviator's Certificate No. 1; the Navy had its first official pilot.

Although the US Navy was the first to operate aircraft from ships, the British Navy had taken an interest in air matters from an earlier date. An awareness that aviation could play a big part in future wars grew with the massive 'arms race' developing between Britain, France and Germany. In 1909 Blériot flew the English Channel, indicating that Britain could no longer consider herself immune from assault from the air. In the same year the first big Zeppelin, Z1, went through its paces in Germany, clearly showing its military potential. It was directly to counter the threat of the Zeppelin that the British Admiralty allocated £35,000 for the development and construction of a giant dirigible airship at Barrow by Vickers-Armstrong. Appointed to supervise its construction was Captain Murray Sueter. He later chose two naval lieutenants, Usborne and Talbot, as pilots for the craft.

Construction of the big airship, designated 'Naval Dirigible

The Caproni Ca series were the
second strategic bombers to see
service, preceded only by the
first Sikorsky machine. Operated
by the Corpo Aernautica
Militare, they flew many
missions on the
Austro-Hungarian front during
the First World War

No. 1' took nearly two years, and the eventual cost worked out at £41,000. At about 420 ft it was slightly shorter than Z1 and her successors. It had twin Wolseley engines giving a total output of 400 hp, a gas capacity (hydrogen) of 700,000 cu ft, and a lift of 21 tons. There were two gondolas beneath the envelope for the crew, connected by a catwalk. The two engines were on outriggers abreast the leading gondola. There was a further small engine driving a pusher propeller aft of the rear gondola. Quadruple rudders and triple elevators were carried on the tail fins, and there was a secondary rudder beneath the envelope and a further elevator under the nose.

This 'leviathan of the air' had a designed top speed of 40 mph and weight was controlled by water ballast. It was to be moored to a tower on a floating raft, and the idea was that the raft could be towed around with the fleet by a suitably equipped warship. The old light cruiser HMS *Hermione* was assigned to this task, the first British ship specifically earmarked for aviation duties.

On 22 May 1911, the launching took place, and Naval Dirigible No. 1, named *Mayfly*, was eased out of its shed built over one of the Vickers slipways, and 300 sailors from HMS *Hermione* hauled and pushed the big craft out over the waters of the Furness. Trimmed to ride just above the water, and aided by a motor launch between the gondolas, *Mayfly* was taken to the mooring tower in nearby Cavendish Dock. Launched at just after 4 o'clock in the morning, it took about an hour to moor; it was the only successful journey.

After testing the efficacy of the mooring arrangements, *Mayfly* was taken back into the shed for 'fitting out'. By 23 September 1911, after tests of the engines and controls in the shed, Captain Sueter formally accepted *Mayfly* on behalf of the Admiralty. Next morning, the 24th, the sailors from HMS *Hermione* once more mustered at the shed to ease the craft out for a proper maiden flight. As it was hauled out tail first, a tug stood by to turn *Mayfly* about and it was while doing this that disaster occurred. As the craft was hauled round, it started to cant alarmingly to starboard. Efforts were made to correct the trim but there followed a loud tearing and cracking and *Mayfly* cracked in half amidships and dropped and settled into the water with the tail floating skywards. The crews in the gondolas swum for their lives, but fortunately there were no fatalities, though Naval pride took a hammering and wags promptly redubbed the craft 'Won't fly'.

It marked an end of ambitions to build airships matching those of Germany and, indeed, in the two years since the project was conceived in 1909, the Germans themselves, not to say other nations, had experienced similar set-backs with very large airships. Opinion now changed in favour of smaller and handier airships; the Royal Navy was later involved in several projects of this nature and eventually mustered a large and useful airship fleet in the First World War.

Ely's achievements in taking off and landing from the *Pennsylvania* (and earlier the *Birmingham*) in America, had a big influence on British ideas in this direction. The magazine *Flight* commented soon afterwards, 'The recent demonstration by Mr. Eugene B. Ely of the possibility of aeroplanes working in conjunction with the Naval arm of the services has brought home to the (British) authorities more rapidly the importance

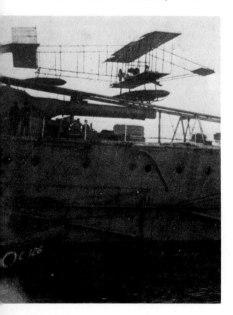

A Short S.27 being taken aboard before being launched from HMS *Africa* in January 1912

of flying machines as an auxiliary for scouting purposes than the most sanguine enthusiasts could have hoped for.'

In November 1910, soon after Ely's first flight from USS *Birmingham*, Admiral Sir George Neville, naval commander in the Medway, arranged a series of lectures on aviation for officers under his command. At these lectures Navy and Marine officers were invited to volunteer for flying courses – on condition that they had not already specialised in some other arm and were not married. From some 200 volunteers, the British Admiralty chose four for the first course. These were Lieuts. Samson, Longmore and Gregory, and Captain Gerrard of the Royal Marines. Since the Royal Navy did not then, of course, own any aircraft, the officers were sent to Eastchurch airfield on the Isle of Sheppey where Short Bros, the pioneer British plane makers, had their factory, and where the Royal Aero Club, Britain's premier flying organisation, had its headquarters and hangars. An instructor from the club, using club aircraft (Short S.27s which were based on the French Farman design), volunteered to teach the four officers, who duly started their course in March 1911.

By summer that same year they had all qualified and became instructors for subsequent batches of volunteer pilots. By then, also, Admiral Prince Louis of Battenburg had succeeded Admiral Neville and enthusiastically espoused the aviators' cause. He persuaded the Admiralty to purchase the Short S.27s and order a few more aircraft of different types. Prince Louis' 11-year-old son was taken for a few flights, a not inappropriate introduction to aircraft for the boy who was later to become Admiral of the Fleet Earl Mountbatten and a great believer in naval air power in the Second World War and post-war years.

Until now, the fledgling Navy fliers had all been firmly landbased but in September 1911 Commander Oliver Schwann, one of the later group of pilots to qualify, took a newly-purchased Avro D Type aircraft to Cavendish Dock, Barrow-in-Furness, fitted it with floats of his own design and, basing himself and the Avro on HMS *Hermione*, carried out a series of taxying trials, made some adjustments and improvements, and in October 1911 succeeded in taking off from the water for the first time. Thus, for a short period, HMS *Hermione* became the first British seaplane tender.

Prince Louis of Battenburg was instrumental in getting Admiralty authority for the conversion of two obsolete pre-Dreadnought battleships with flying-off platforms for Naval aircraft experiments. In November 1911 work was put in hand in fitting wooden stages on the bows of HMS *Africa* and HMS

Igor Sikorsky's *Ilya Mourometz* which was developed to become the world's first heavy bomber, capable of carrying a bomb load of 2,200 lb in the later Ye variant. More than 75 of these giant four-engined aircraft were built

Hibernia following the same sort of layout as had been used on USS *Birmingham* a year earlier. Meanwhile, Shorts had co-operated with the Navy fliers in adapting inflatable rubber air bags to fit to the undercarriage of the S.27 training machine. Lieut. Arthur Longmore successfully tested the new float on 1 December 1911, taking off from Eastchurch and alighting on the River Medway.

By the first week of January 1912, the improvised wooden flying-off platform had been completed on the forecastle of HMS *Africa* and the Short S.27 (No. 38), fitted with the air-bag floats which Longmore had tested, made a successful take-off piloted by Lieut. Samson.

It was in April 1912 that service aviation in Britain was put on a formal basis, when plans were announced for the establishment of a Royal Flying Corps to come into being on 13th May 1912. This new formation was to administer and take charge of the existing Army Air Battalion—which became the Military Wing—and the Navy fliers who would form the Naval Wing. The new Naval Wing's headquarters were to remain at Eastchurch and the existing establishment there was given the formal title of Naval Flying School.

The government white paper on the subject of the RFC stated: 'It is impossible to overestimate the importance of experiments for the development of hydro-aeroplanes, and in flying from and alighting on board ship and in the water under varying weather conditions. Until such experiments have proved conclusively how far such operations are predictable it is impossible to forecast what the role of aeroplanes will be in naval warfare, or to elaborate any permanent organisation. The present organisation must therefore be regarded as provisional.'

On the question of equipment it mentioned that 12 new aeroplanes and hydro-aeroplanes, plus the necessary floats, were to be purchased, and stated that 'it has been deemed desirable to test the number of types with a view to arriving at the most satisfactory for naval service'.

Experience with the *Mayfly* had left its mark. The white paper explained cautiously, 'The prospects of the successful employment of the rigid type of airship are not sufficiently favourable to justify the great cost and it is therefore recommended that the Naval experiments should be confined to the development of aeroplanes . . . The utmost vigilance will be taken, however, in watching foreign developments of the airship and the present recommendation will not be taken to prejudice a reopening of the question should important developments occur.'

The Naval Wing of the RFC made a spectacular name for itself from the very beginning. On 9 May 1912, King George V was to review the British Fleet at Weymouth and the Navy fliers were to participate in fleet manoeuvres for the first time. A temporary seaplane base with tented hangars was set up on a slipway at Portland Naval Base and as the royal yacht *Victoria and Albert* approached Weymouth, the four pioneer Naval aviators, Samson, Longmore, Gregory and Gerrard, set out to find the vessel from the air in misty conditions. Samson found it first and circled the ship at little more than masthead height, followed later by the other three, before they all flew the 12 miles back to Portland.

Having effectively demonstrated their 'scouting' ability, the first British demonstration of the potential offensive use of Naval aircraft took place the same afternoon, when Lieut. Gregory dropped a dummy 300 lb bomb from a height of 500 ft within a short distance of the *Victoria and Albert*, which was by this time at anchor in Weymouth Bay. Spotting a submerged submarine at periscope depth, Gregory provided another exciting spectacle for the watchers on the royal yacht by swooping low over it at only 20 ft in a simulated bombing run. While all this was happening, Commander Samson had taken off from Portland with a courier aboard carrying a letter for the king. He put his Short S.41 seaplane into the sea alongside the yacht who collected the courier by boat, after which Samson returned to base. The forceful and extrovert Samson, who had organised this impressive show of Naval air power, also persuaded a few friends from the Royal Aero Club to bring their aircraft along, so that with aviation activity taking place all day the Naval fliers virtually stole all the news headlines from the ships of the fleet.

On the following day, however, an even more spectacular event took place when Commander Samson flew the Short aircraft No. 38 from HMS *Hibernia* while she was steaming at 15 knots towards Weymouth, thus becoming the first Naval aviator ever to take off from a ship under way. Though a considerably more hazardous feat than taking off from a ship at anchor, the actual launch was very much more satisfactory than previous take-offs at anchor, due to the added wind speed from the moving vessel. The aircraft lifted swiftly and easily from the deck, gained height quickly, and flew ashore to the landing ground at Lodmore, near Portland.

History was made and the Admiralty was so impressed that funds and encouragement for the expansion of the Naval Wing were swiftly forthcoming. And for the time being–largely because war clouds were looming in Europe–the British Navy had caught up and passed the US Navy in the field of naval aviation.

Royal Flying Corps

In Britain an autonomous arm was formed in April 1912, the Royal Flying Corps, to undertake responsibility for military aviation. Military aircraft trials were held on Salisbury Plain for the War Office to decide the best sort of aircraft for the new corps to use. The trial, and a £4,000 prize was won by Colonel Cody with a design of his own which competed against established Farman and Voisin types. Cody's aircraft was a very large 'box-kite' type machine which was nicknamed the 'Cathedral'. It crashed the following year, however, killing Cody and his passenger. A policy was initiated of using aircraft types built exclusively by the Royal Aircraft Establishment at Farnborough. Thus came the various types such as the BE (Bomber Experimental), FE (Fighter Experimental) and RE (Reconnaisance Experimental). Several models in these categories were produced, and for the most part they were successful. However, when war came in 1914, with a need for rapid expansion, there was a reversion to buying aircraft from commercial companies. A young designer called Geoffrey de Havilland, later to become famous with his own company, designed the various BE models.

One odd anomaly was the incorporation of the pioneer naval flyers into the Royal Flying Corps as the 'Naval Wing', but after a short period of protest the Royal Naval Air Service (RNAS) was set up. In practice this caused much duplication of effort since there were now, in effect, two air forces in England. In 1918, however, the RFC and RNAS were merged into one arm, the Royal Air Force.

Stunts and Races

There were many exciting developments in the world of aviation at this time. Air races and tours by famous flyers were capturing the imagination of the public as never before, and some practical, commercial ideas for exploiting the use of aircraft were being realised. One of the first of these, and pre-eminent ever since, was the airmail service. The first official airmail flight was made in June 1911, commemorating the coronation of King George V, a suitably 'modern' idea and establishing in the public eye that aircraft were here to stay. That first airmail route was a short one—from Hendon to Windsor. In September 1911 the

A Short 184 seaplane

first airmail flight in America took place, and thereafter a number of such services were established in the major nations of the world.

The dividing line between stunts and practical achievements was a thin one. At one of the 1911 Hendon shows, Claude Graham-White staged air attacks on dummy forts and battleships, using sacks of flour as bombs – great entertainment, but it also showed the way events would go only three or four years later when the major air forces would start dropping real bombs on real targets. Aerobatics, another feature of the new-fangled air shows, were the proving ground for the manoeuvring and handling skills in the 'dogfights' which were to become such a feature of the air war in Europe from 1914.

One dramatic event in October 1911 was the first ever coast-to-coast flight in the United States. The newspaper magnate William Randolph Hearst offered $50,000 for the first man to cross America within 30 days. Using a Wright aircraft, a racing driver, Calbraith Rogers, flew 4,000 miles from New York to Pasadena – but in seven weeks rather than 30 days. Sponsored by a soft drinks manufacturer, the flight attracted some brash publicity which tended to overshadow its importance as an aviation milestone.

Other newspaper magnates did much to sponsor flying events and this furthered the popularity of aviation in the public eye. James Gordon Bennett, owner of the *New York Herald*, started a sponsored balloon race in 1909, and thereafter sponsored a whole series of Gordon Bennett Races for aircraft, in one of which Claude Graham-White carried off the honours for Great Britain.

The 1912 Gordon Bennett races at Chicago was won by Frenchman Jules Vedrines in a Deperdussin-Béchereau, an advanced, streamlined monoplane with plywood, monocoque fuselage. It was the first aircraft to exceed 100 mph, attaining 107 mph in the race. The next year it managed 125 mph. This was also the first aircraft to have the type of stick and 'wheel' control which became almost universal in subsequent years.

Northcliffe, owner of the *Daily Mail* and many other publications, was similarly famous for the number of races he

The Sopwith Pup (here with a Bristol Fighter flying above it) was one of the best British fighters of 1916–18 – simple, reliable and with a good performance. It was used by the RFC and RNAS

sponsored with large cash prizes. In 1913 a wealthy French speed enthusiast, Jacques Schneider, put up the trophy which bore his name and the cash prizes for a series of seaplane races. A floatplane version of the Deperdussin-Bécherau won the first-ever contest at 45 mph.

By 1913 there was a new generation of designs coming along, some of which saw important service in the war. In a way the air shows, races, and trials were a training ground for war, and some of the planes produced were destined either to play a part or open the way to better developments. Typical was the Sopwith company in England, whose founder T. O. M. Sopwith was a pioneer pilot. His Sopwith Tabloid of 1913 was sensationally original. It set the style for most fighter designs in the 1914–18 period – the classic biplane layout. The Tabloid was originally built as a seaplane for the 1914 Schneider Trophy races, but became the forerunner of a series of Sopwith aircraft which were to become household names in the First World War. Other British firms making big advances just prior to the outbreak of war were Avro, who had by now produced a very successful tractor biplane, Bristol, and Handley Page, who favoured monoplane designs, and Shorts, who were the leading seaplane builders and were the first to introduce stressed metal wing coverings in place of fabric. In Germany, a very fast and successful tractor biplane, the L.V.G. had achieved a record for a non-stop, 24 hours flight in 1913, and in Holland a young designer, Anthony Fokker, had attracted some attention and his services were sought by the embryo German Army air service.

Most spectacular of all aircraft in the 1913–14 period, however, were the huge, four-engine machines made by a young Russian designer, Igor Sikorsky. He reasoned that large, multi-engine aircraft were needed to obtain the necessary power-to-weight for a satisfactory performance in the air. In 1913 he built what is believed to be the world's first four-engined aircraft, a one-off prototype called 'The Grand'. In 1914 he produced an improved version, the 'Ilia Mourametz'. This machine flew between Petrograd and Kiev and back, around 1,600 miles. The aircraft also held height and payload records and was very much a progenitor of the large, multi-engine bombers which would be built in ensuing years.

War in Europe

Events of August 1914 soon swept Britain, France and (from 1917) the United States into the war in Europe.

Air forces in August 1914 were still very small and rudimentary in their equipment. The British Royal Flying Corps at this time, for instance, had only 105 officers and 63 aircraft. The French had well over 200 aircraft and the German Military Aviation Service had almost 250 assorted aircraft, about half of which were Rumpler Taube monoplanes, the most successful of the pre-war German military designs. In the United States the US Army Signal Corps was now somewhat behind in establishing an air branch, but it did so in 1913 with a few aircraft purchased quietly from Wright, Curtiss and Martin.

Very little official thought had been given to the aircraft's role as a warplane, and many of the functions developed came

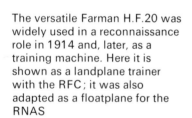

The versatile Farman H.F.20 was widely used in a reconnaissance role in 1914 and, later, as a training machine. Here it is shown as a landplane trainer with the RFC; it was also adapted as a floatplane for the RNAS

An F.E.2d in typical heavily armed form

about as the result of actual combat experience. The aircraft was largely regarded as another pair of eyes for the soldiers on the ground, and in this reconnaissance role aircraft reigned supreme throughout the war. On the Western Front in the early weeks of the war aircraft from both sides made discreet flights to report back on enemy dispositions. There was of course no radio, and messages about enemy troops or to direct gun fire were dropped in containers or signalled by other means. The tethered balloon came into its own once more as a useful but fairly static means of observation.

The war was only a week or two old when the first shots were fired in the air. Initially, none of the aircraft in service with the French, British, or Germans were armed, but the pilots and observers carried pistols or rifles and took shots at opposing aircraft. By the time of the Battle of Mons in September 1914, when the small British Expeditionary Force succeeded in slowing the powerful German advance into Belgium, the British C-in-C was able to record in official despatches that aircraft of the Royal Flying Corps had 'succeeded in destroying five of the enemy's machines'.

Most of the machines in use by the air forces in the opening months of the war were famous types from peacetime years, albeit usually in modified or developed form. Thus Henry and Maurice Farman types were to be seen; the latter, the Maurice Farman 'Shorthorn', was subsequently used extensively as a successful and forgiving training aircraft for the thousands of young men who wanted to fly. The Blériot, Bristol Scout, Nieuport Scout, and Avro 504K (again a type later more famous as a trainer) were also used by the Allies, as were the R.E.8 and B.E.2, products of Farnborough specially built for the RFC. On the German side early, unarmed, 'scout' types were the Rumpler monoplane, Albatros and L.V.G.

By the end of 1914 a more colourful development had been the almost universal adoption of national markings to enable friend and foe to recognise each other, and to avoid unnecessary incidents involving the shooting down of 'friendly' aircraft. The Germans adopted their national patée cross, the French, a tri-colour cockade, and the British, the Union Jack. These markings were quickly developed further. First to change were

51

the British. The configuration of the Union Jack looked too similar to the German cross at a distance or in poor light, and the British therefore adopted the French tri-colour cockade, but reversed the order of colour to blue, white, red, the latter in the centre. Subsequently, other allied nations also used cockades in their national colours.

Originally, military aircraft were just like any other machines of the time. The fabric-covered areas were unpainted – just varnished. First came the national markings, then some type of identifying numbers, squadron codes or serials. Slogans or personal markings appeared, painted lovingly and artistically on to the machines by their air or ground crews. Aircraft parked on the ground were conspicuous, however, when finished only in plain, light cream, fabric covering. Soon camouflage paints were applied to upper surfaces, usually browns or greens. The Germans used plain camouflage colours but later made very wide use of lozenge pattern covering – printed, multi-hued fabric which broke up the outline of the aircraft to some extent by the 'honeycomb' pattern of colours.

As the war progressed a great many pilots became 'aces' through their prowess at shooting down unusually large numbers of enemy aircraft. Such ace pilots sometimes painted their aircraft in flamboyant colours. Perhaps the most famous was the Red Baron, whose aircraft was alleged to be painted all red, and was certainly painted partly red. In the event, the subject of aircraft markings and colours caught the imagination of many youngsters and air enthusiasts, and has remained an interesting line of study ever since, so diverse have been the styles and applications – a veritable heraldry of the air.

It did not take long for the airmen to realise that armed aircraft were needed for the successful destruction of the enemy, not the crudity of hand guns. Clearly the facility for firing forward in the direction the aircraft was flying was a major advantage, for the pilot was able to aim his gun – if of the fixed type – by simply pointing his aircraft at the target. The aircraft able to fire forward also had a distinct tactical advantage over one with its gun mounted to fire to the rear or in a limited arc. A pusher-type aircraft could have an observer right forward with a flexible gun mount and able to select his target. The British developed several two-seater pusher fighters, all of generally similar layout – the Vickers Gunbus and de Havilland

below, right
The Morane Saulnier Type-L, or M.S.3, was an advanced design in service in 1913, and was used by the French and British. In 1915 steel deflector plates were mounted on the propeller (visible here) to enable a cowl-mounted Hotchkiss machine gun to fire forward through the propeller arc. Aviation Militaire ace Roland Garros, posed in front of his aircraft in this photograph, used this to great effect

below
The Sopwith Triplane. This was a highly manoeuvrable scout, which led several German companies to start work on triplane designs

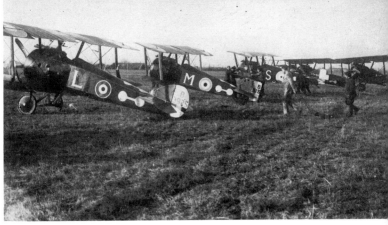

D.H.1, D.H.2, and the F.E.8. The D.H.2 was actually a single seater and something of a death trap, being hard to control while the pilot handled the gun as well; it was known as the 'spinning incinerator'. One of the problems with the pusher aircraft was the quite limited arc of fire for the forward-mounted machine gun and the generally less agile performance of this kind of aircraft. Nonetheless, some of them were well liked and the Vickers Gunbus had a great reputation. One of them shot down the German ace Immelmann.

The most manoeuvrable and fastest aircraft were the single-seater tractor machines and some early, successful efforts with these soon solved the problem of forward-firing guns and the propellers which revolved in front of them. Credit for being the first combat aircraft able to fire forward through its propeller arc goes to the French firm of Morane-Saulnier, whose 'monoplane de chasse' was in service by April 1915.

Developed by Roland Garros, a famous French racing pilot of pre-war days, the idea was simple. The Morane was a high-wing monoplane with the machine gun mounted above the engine cowling. A machine gun mounted thus would in normal circumstances simply shatter the propeller blades. Garros overcame this problem by fitting metal deflector plates to the propeller plates so that any bullets striking the propellers were deflected, but sufficient got through the arc to give a stream of fire. Initially, in service in March 1915, the Morane proved a great success until the Germans discovered the trick and learned to avoid its approach. Late in April 1915 a Morane force-landed and the Germans discovered its secret.

As a result of this, the Dutch designer Anthony Fokker was asked to produce something similar or better. Fokker came up with his famous interrupter gear which in turn was built in to the Fokker E.III Eindecker (monoplane). In service from October 1915, the Eindecker reigned supreme. This was the age of the 'Fokker Scourge' when the continuous forward fire through the rotating propeller gave a rate of fire superior to that of any enemy machine. Fokkers dominated the air fighting scene until about May 1916, when a new generation of British fighters began to appear, types such as the S.E.5 and S.E.5a, the Sopwith Camel, and the Sopwith Pup. The British fighters, with forward-firing fuselage guns, had the Constantinescu synchronised equipment, which was technically superior to Fokker's interrupter gear.

In fighters, once the problem of forward firing was solved, speed and manoeuvrability became important, as did rate of climb and turn. The fighter was now able to escort reconnaissance (or, later, bomber) squadrons over the enemy lines, or carry out independent patrols. The enemy was similarly engaged and when rival patrols met, a terrific 'dog fight' or

above
The Sopwith Camel, developed from the Pup, was the most formidable British single-seat fighter in 1918

above, left
A B.E.2c of the Royal Flying Corps in 1915

53

above
A Spad XIII in the colours used
by Eddie Rinkenbacker. It was
built by Curtiss in the USA

left
A Fokker Dr.1 Triplane similar
to the type flown by Von
Richthofen's squadron on the
Western Front in 1917

opposite
Spad VIIs of the famous
Escadrille SPA3 'Les Cicognes'.
This fine combat aircraft saw
service with most Allied air
forces

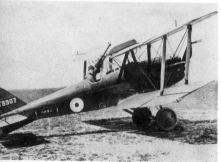

above
An early production S.E.5 fighter showing its Lewis gun mounting

top
An ace pilot and classic fighter of the First World War; Captain 'Billy' Bishop VC and his Nieuport Scout in 1917

melee ensued. These were the days, 1916–18, of the classic small biplane or monoplane fighter and the days when names were made and lives were lost among the daring young men who flew the machines. Some of them, no more than 18 or 19, then became the glamorous heroes of the age. MacCudden, Ball, Bishop, Von Richthofen, Udet, and Immelmann, were all among the best known and the most highly decorated. The skill lay in aircraft handling and being able to outrun the enemy. The Immelmann turn, a half loop with a half roll on the top, was one of the best-known ruses of this period, a manoeuvre which could put a disadvantaged pilot back into a commanding position above his attacker.

In the First World War, too, the concept of the long-range bomber was conceived and proved practical. The tactical bomber had already come into being; several of the different types of recce aircraft were equipped to drop bombs. Originally they were dropped by hand, but light bombs which could be mounted under the wings (up to 25 lb) were also developed. As has already been recounted, the Russian, Igor Sikorsky, had already designed and built a large, multi-engine machine–the Ilia Mourametz. In 1915 the Russians had a number of improved aircraft of this type in service fitted as bombers. The wingspan exceeded 100 ft and 1,000 lb of bombs could be carried. There was also a tail gunner's position, and in general layout and disposition the aircraft set the pattern which most other designers followed. The size and multi-engine layout was dictated by the need for long range to penetrate well beyond the enemy lines.

Other nations soon produced heavy, long-range bombers of their own. In Britain the Handley Page firm worked through most of 1915 to build their spectacular big bomber, the 0/100 for the RNAS, which was finally ready in prototype form by December 1915. There were troubles with the engines which failed to give sufficient power and caused extensive drag due to the large radiators. Changing the radiators and extending the control surfaces helped matters, but sceptics in high places doubted that such a large and unwieldy aircraft could ever fly successfully. Largely due to the forcefulness of Commodore Murray Sueter, head of the Admiralty Air Department, the many problems were overcome, though not without extensive redesign work which included replacing the original enclosed cockpit with an open one. The 0/100 was subsequently developed into the 0/400, an enlarged version with met an Air Ministry requirement for a bomber with a 3,000 lb bomb load for night attacks on Germany. An even bigger development was the four-engined V/1500, intended to be bigger than anything the Germans were planning, though this type never saw combat before the war ended.

The 0/400, however, did see extensive and successful service. Strategic bombing was initially the brainchild of the RNAS, who had asked for the 0/100 as a heavy bomb carrier after the first squadrons of the Royal Naval Air Service in France had shown what could be achieved by using their tiny Sopwith Tabloids as bombers in attacks on Zeppelin bases at Dusseldorf in October 1914. German Zeppelins, and later heavy bombers, attacked London in the middle of 1917. At this the Air Ministry took over the idea of concentrated bombing of Germany. One long-term result of this was the unification of the Royal Naval Air Service and the Royal Flying Corps into a single Royal Air Force. The commander of this force was Major General Hugh Trenchard, the last commander of the RFC in France before the RAF was formed. Trenchard pushed the idea of long-range bombing attacks on Germany; in the event, Germany sued for peace in November 1918, before his plans could be got under way on a big scale. Trenchard became the true architect of Britain's Royal Air Force, which was to set the pace in techniques and ideas in the 1920s and became the envy of the world in the quality of its training and the perfection of its flying skills.

The 0/400 was an important aircraft in other ways. As we shall see, it was big enough to be used as an airliner when peace returned. But it had (and the 0/100 and some other types) the new Rolls-Royce Eagle engines; the Eagle was a very reliable and powerful (360 hp), V6, water-cooled unit which was a huge step forward in aero-engine design. It was the progenitor of a famous series of Rolls-Royce engines (all named after birds of

The mighty Handley Page 0/400 bomber had folding wings to facilitate hangar storage

prey) which set the standard by which all other engines were measured.

It was the Germans who made the early running with strategic bombing, however. They started bombing London with their huge naval Zeppelin airships with some success–haphazard but a powerful psychological blow against the civilian populace of Britain. But British fighter pilots, using techniques of great daring and skill, gradually got the measure of the Zeppelin menace, and German losses were high. The legendary Gotha bomber–there was actually a series of them–took over where the Zeppelins left off. The G-IV and G-V were the most widely built and used, and from 1917 they carried out raids on the Western Front, Paris and London. A tail gunner, with a very good, all-round field of fire (achieved by using a 'tunnel' under the fuselage) was an outstanding feature. The aircraft were plywood covered and very sturdy and there was a well-sprung undercarriage. Mercedes 260 hp engines provided the power and the span was $77\frac{3}{4}$ ft, and length 39 ft. Because of their high operational ceilings (15,000 ft) they were very successful on the initial bombing offensive against England, in summer 1917. The British fighters took so long to reach this height that the bombers had released their loads and headed for home before they could be intercepted. New British fighters, such as the Sopwith Camel and Bristol Fighter, were soon in service and were able to deal with the Gothas more effectively, so much so that they were confined to night bombing in 1918. The Friedrichshafen GIII was the Gotha's contemporary and was used with equal success.

The United States were late entrants to the war in Europe, and in April 1917 when they became involved the US Signal Corp (then responsible for army aviation) had only 55 aircraft, while the US Navy had 54. A huge programme of expansion was started, with the US forces initially adopting existing British and French designs. The French Nieuport and Spads of various types equipped the American pursuit squadrons. A fine 'standard' aero engine was designed, the Liberty V8, liquid-cooled, 400 hp unit, and was in production before the war ended. A British fighter-bomber, the D.H.4, was selected as a standard design for the US Army, fitted with the Liberty engine and built under licence in America. An important home-built type which became almost 'immortal' was the Curtiss JN-2 'Jenny' trainer, turned out in thousands. Several other US types were designed and produced, though too late for war service. Outstanding was the fine Thomas-Morse Scout, a close copy of the British Sopwith Camel fighter, and the Martin MB-1 heavy bomber, an equivalent type to the Gotha or 0/400.

By 1917–18, in fact, though speeds had not dramatically improved, the quality of design had. Numerous aircraft earned themselves legendary reputations in the air. Of particular importance were the Sopwith fighters, each a logical improvement on the other, the superb Fokker fighters and the classic Albatros series. By 1918, the Germans had shown the way to the all-metal monoplane fighter, with the Junkers J.10 and J.11, but the war was over before these advanced types were in service. The Junkers J.4 biplane was in service however, all metal and well armoured. It was the first, purpose-built, ground attack aircraft and pointed a long way ahead to the next Armageddon.

opposite, top
Flight Commander Dunning successfully landing a Sopwith Pup on the forward deck of HMS *Furious* in August 1917. HMS *Furious* was later rebuilt as an aircraft carrier with a full-length flying deck

opposite, lower
The Fokker DVII, the finest German fighter in service in 1918

below
The French Hanriot HD–1 was influenced by the Sopwith designs. This aircraft is in Italian service in 1918

The Infant Airlines

The first, commercial, passenger-carrying craft to operate what would now be called scheduled services were, in fact, airships. The idea of carrying passengers in the air on a grand scale goes back to Henson and others who were active in the nineteenth century. Probably the first airline route of all was between Lake Constance and Berlin, which was flown by Zeppelins, starting in March 1911 and only cut short as a continuing operation by the outbreak of war in August 1914, by which time over 19,000 passengers had been carried.

First of the passenger-carrying Zeppelins was the 'Schwaben', which could seat about 32 people. Count von Zeppelin's company, the Zeppelin Transportation Co., actually carried over 35,000 paying customers when other, lesser routes were counted up to August 1914.

The early heavier-than-air aviators soon cottoned on to the idea of carrying passengers, and 'joy riding' became a feature of some of the first air pageants, an exciting adventure for those daring (or foolhardy) enough to take advantage of the opportunity. One of the first Breguet biplanes built, which had a 100 hp liquid-cooled engine, made an experimental flight in 1911 with 10 passengers and showed the potential, though the aircraft was not designed or intended as a passenger carrier.

In 1912 the Avro Tandem appeared, strictly speaking a light aircraft, but it had the distinction of a fully enclosed cabin for the pilot and his passenger and established the layout commonly used years later for passenger-carrying light aircraft. Also in Britain, the enterprising Claude Graham-White built a special passenger carrier, the 'Charabanc', which was a pusher type with larger than usual nacelle which held up to 10 passengers plus the pilot. The aircraft had a 100 hp, six-cylinder, in-line engine, and earned its living from joy rides at Hendon in 1913.

A company in the United States started the first recorded scheduled service by aircraft. This was the St. Petersburg-Tampa Air Boat Line which ran a twice-daily service between these two points, across Tampa Bay, Florida, limited by the fact that their Benoist flying boat could only carry a single passenger on each flight.

It took the impetus of war to bring about passenger flying by aircraft as a practical proposition. The important factors were the development of longer ranged, and more reliable aircraft (the bombers and light bombers), the availability of a supply of these aircraft after the war, and most important, a trained supply of pilots and mechanics who were able to provide the essential manpower and knowledge for post-war operations.

opposite, lower
The early days of French civil aviation. Potez 9s at Le Bourget in 1921

Also to prove important later on was the development of the all-metal aircraft (notably by the German Junkers firm). While the first Junkers were warplanes, the principles were established and within a decade or so, big, all-metal passenger aircraft were flying and were eventually able to replace the wood and fabric airliners of the first generation.

During the later years of the war, 'courier' flights between London and France had been started for officers and officials, and it showed how feasible a London-Paris air service could be. The first British flight, from Paris to Kenley, Surrey, took place on 8 February 1919, with military men as passengers, since

below, left
Passengers embarking in a Sablatnig P111 of Lloyd Luftverkehr Sablatnig Airlines in 1919

below, right
The Breguet 14 was a French bomber converted to one of the earliest airliners

The Farman Goliath used on the first Paris–London flight in February 1919

commercial flying was still forbidden under war regulations. The aircraft was a Farman Goliath. The first Paris to Brussels flight of a similar nature was flown by a Caudron C.23 only two days later—these were the start of what is now an intricate and complex network of air routes all over Europe.

Later in 1919, as soon as restrictions were lifted, there was a rush to set up small airlines, often by ex-service pilots who acquired one or two (sometimes more) ex-service machines. Many of these firms led a precarious life, and bankruptcies and takeovers were common. The four biggest airlines operating from England to France by the early 1920s were Instone Air Lines, Daimler Hire, British Marine Air Navigation, and Handley Page Air Transport. The latter had converted some of their 0/400 heavy bombers to become the first, big, multi-engine air liners. A typical small operator in 1920 was Air Transport and Travel Ltd, who had a small fleet of former RAF D.H.9 light bombers. The D.H.9 and its predecessor, the D.H.4 , were popular choices for airline work, being cheap to run and easy to convert. Some were given enclosed or partly-enclosed cabins behind the pilot's station.

By 1924 even the leading four operators in Britain were finding financial conditions too harsh for them to remain in business, and on a suggestion from a committee set up by Winston Churchill, then Minister for Air, the four companies and their aircraft were merged to form a single national airline, Imperial Airways, progenitor of what is now known as British Airways. There was a similar sequence of events in most other countries through the 1920s, though KLM, the Dutch airline, was set up in 1919 as a national airline from the start. Air France, on the other hand, was not formed (from amalgamation) until 1933. Germany, as the conquered state, was not originally allowed to have an airline and German efforts were thus concentrated on gliders, a field in which that nation became pre-eminent between the two world wars. It is also of passing interest to note that the embryo Luftwaffe–the new German Air Force of the mid 1930s–was trained up in its initial stages almost entirely with gliders. German aircraft designers kept themselves in business, and up to date in techniques, by working with friendly foreign manufacturers until 1925, when they were at last freed again to built aircraft in Germany.

In the USA things were slightly different. Distances were much vaster and there was no great and tempting sale of ex-service aircraft on the scale found in Europe. So the first air routes were mostly sponsored by government agencies, such as the Post Office. The US Post Office set up an air route from New York to Philadelphia in 1918 specifically to carry mail.

Curtiss Jennys of the US Army were initially used. Then similar routes were started out of New York in other directions. Civil pilots and ex US Army D.H.4s were later employed. Subsequently, a new breed of aircraft appeared, the purpose-designed 'mailplane', initially with a mail compartment, then with the facility for carrying a few passengers in an enclosed cabin. The first ever Boeing design–the B-1 'Mail Boat'–was a small, Liberty-engined, biplane flying boat with a plywood hull specially for mail service. In 1921 a developed version of the Martin MB-1 bomber appeared–the Martin NBS-1–for US Army service as a bomber, but in alternative versions as a mailplane and a 12-seat passenger liner. These

were but two of several assorted designs of the era. Night flying and 'blind' flying in bad weather were skills which had to be mastered by the mailplane pilots—despite the lack of instrumentation in the early 1920s—for the US Mail had a reputation for getting the mail through on time. Light beacons were subsequently used extensively to guide the mailplane pilots along their routes at night, and this marked the beginning of the idea of using navigation aids from the ground.

The early scheduled routes, such as Brussels-Paris, London-Paris, and New York-Philadelphia, seem to be short 'hops' by modern standards. In the early 1920s, however, any such journey was a great adventure. London-Paris, for example, was theoretically a less than three-hour journey, but there were instances of a single journey taking over 24 hours. The early

top
A D.H.4a of the RAF which was used for Paris-Kenley courier flights in 1919

above
The formal ceremony marking the formation of Air France from smaller airlines, at Le Bourget in October 1933. The aircraft, from left to right, are: Wibault 282 T12, Farman F.303, Latécoère Laté-28, and Loiré et Olivier LeO 213

A Handley Page 0/400 of
Handley Page Air Transport sets
off for Paris in 1921

airline pilot in, say, a converted Handley Page 0/400 bomber, had to be a master of all trades. His crew companion was a flight engineer and between them they had to be prepared for almost any eventuality. The engines of the aircraft were reliable for their day, but they could still be temperamental. Forced landings were thus not unusual. In fact, they were to some extent expected and Imperial Airways went to the trouble of providing the pilot with rail timetables and routes to the nearest Channel ports so that passengers–and the mail–could continue the journey by train and packet boat from wherever the aircraft landed. In Europe as in USA, the airmail was a novelty, and it was more important to the airline as an on-going contract than the passengers; that at least is the interpretation from a reading of the early airline regulations. There are several, authentic, recorded instances of the pilot's initiative being used to the full in order to get the mail to its destination; on one occasion an 0/400 was forced to land in France well short of Paris. The pilot hired a horse from the farmer in whose field he had landed, and then rode five miles bareback to the nearest railway station, there to place the mail on a Paris-bound train. Imperial Airways pilots in the mid 1920s carried a £5 'float' at all times to cover such eventualities.

Navigation was extremely primitive by modern standards. Flying over the Channel in mist the pilot would lower his wire aerial (a crude radio was carried) and let it dangle at 50 ft–when the wire was felt to splash on the water surface a rough idea of a minimum safe height was available. Leaving the original Imperial Airways airfield at Hounslow Heath, the pilot would follow the Thames to its mouth, then turn south to head across the Channel. If the track of the river was not visible–in bad weather–an alternative would be to follow the railway lines, sometimes with a low pass to read the nameboard of the station! Navigation, in fact, had hardly made any progress since the days of the Wrights, Santos-Dumont, and Farman, before the war. Air traffic control was not so much primitive as almost non-existent. The regulations for aircraft using the main Paris

An Imperial Airways Argosy of 1926

airport in 1924 were short and simple: when more than one aircraft was in the air inside the airfield boundary, the rules merely stated that 'the aircraft nearest the ground shall have precedence for landing'.

Early air passengers were relatively few and far between. They had to be quite well off. The London-Paris fare was £15 15s, a large sum for the early 1920s. There was no cabin staff and if any messages had to be passed in mid-air to the passengers, it was the flight engineer who had to crawl back from his open cockpit to impart the information. The passengers – from two to twelve, depending on the aircraft – travelled in wicker seats or on flimsy wood benches. In case a passenger needed a meal in flight in the early days, he was enjoined to 'tell the pilot before take-off so that a hamper may be put aboard the aircraft'!

In 1919 the great bogey was distance. The long-range bomber, which had just been developed, were clearly also going to be useful in the development of airliners. The British *Daily Mail* newspaper had put up a prize of £10,000 in 1913 for the first aircraft to cross the Atlantic. At this time the longest flight ever was of 630 miles, while the Atlantic crossing was about 2,000 miles at its shortest. While the war was in progress nobody had time to chase the tempting prize, but straight after the Armistice there were a number of contenders preparing for the attempt. An early competitor was an English pilot John Porte from White and Thompson and Co Ltd, which imported Glenn Curtiss seaplanes to England. Just before the outbreak of war Porte went to America to arrange for a Curtiss flying boat 'America' to be ready for a transatlantic crossing attempt. The war started before anything came of this scheme, though Porte shipped the flying boat back to Britain where it became the basis for the type built in England under licence for the RNAS – one of the very first maritime patrol aircraft.

In May 1919 the US Navy made a transatlantic attempt using four specially-designed, four-engined flying boats NC-1 to NC-4 built by Glenn Curtiss. Of these NC-2 was used for spares. The rest left Long Island on 8 May 1919, bound for the Azores. Only

The first Atlantic crossing by seaplane. A Curtiss NC—4 enters Lisbon on 29 May 1919

one machine, NC-4, actually made the Azores by flying. NC-3 forced-landed and taxied the last 200 miles while NC-1 was lost in another forced landing. US Navy ships were strung along the route for safety.

Though this was the first west to east crossing of the Atlantic it was very shortly eclipsed by what at the time seemed an impossible feat, an Atlantic crossing coast to coast by landplane. John Alcock and Arthur Whitten-Brown were the pilots concerned, test pilots for Vickers who had taken a specially-converted Vimy bomber to Newfoundland with them. Even so, they were the second pair to make the attempt. Two other Englishmen, Harry Hawker and Mackenzie Grieve, in a single-engined Sopwith, left before Alcock and Brown but their machine came down in the sea early on, the pilots being rescued.

The Vickers Vimy was another heavy bomber, contemporary of the 0/400, and also powered by two Rolls-Royce Eagle engines. The aircraft was specially prepared for the flight, left unpainted for lightness, and packed with fuel enough for the journey. Leaving St. Johns, Newfoundland, the machine headed out over the Atlantic into indifferent weather. The flight was an epic of endurance. The date was 14–15 June 1919, and the elapsed time of the flight, 16 hours and 20 minutes. Visibility was poor and extreme icing caused the aircraft to come near to ditching in the sea on several occasions. The aircraft finally ended on its nose in an Irish bog (mistaken for a green field). Alcock and Brown were feted, knighted, and received the prize money for their heroic achievement. Alcock died later that year in a crash while testing a new aircraft.

Alcock and Brown's flight was the start of a great period of trail blazing. Routes around the world could be opened up. It was shown that suitable aircraft could be flown almost anywhere. The achievements were prodigious and it is only possible to mention a few of the more notable pioneers. A prize of £10,000 was put up for any Australian who could fly from Britain to Darwin inside 30 days, and two brothers, ex-RFC pilots, Ross and Keith Smith, responded to the challenge. With another Vickers Vimy, and flying via Taranto, Basra, Calcutta and Rangoon, they made good time. Singora (in Siam) and Singapore were more difficult landing grounds. Disaster almost struck at Sourabaya in the Dutch East Indies. A supposedly purpose-built tarmac runway had been built over a swamp. The surface did not bear the aircraft's weight and it sank up to its axles in bog. Local natives made a bamboo runway surface and almost by a miracle the machine took off for the last leg over the Timor Sea to Darwin. They made it in 28 days, just qualifying for the prize. The total flying time was 135 hours and the Smith brothers were the toast of Australia.

To some extent more spectacular was the flight of Raymond Parer, an Australian, and a Scots companion, J. C. McIntosh, in 1920. They acquired a tiny, single-motor D.H.9 to challenge the prize for the first to Australia. While they were testing the machine, the Smith brothers had already arrived at Darwin and claimed the prize. Despite the Smith success Parer and McIntosh set off, and they recorded a number of notable 'firsts'. Their little D.H.9 was the first single-engined aircraft to cross over the Mediterranean, the first single-engined aircraft to fly to Egypt and the first to fly from Palestine to Baghdad across the inhospitable Syrian desert. More records followed for single-engined aircraft. They were the first to Calcutta, and, indeed, the first from England to Australia (as the Vimy was twin-engined). They had trouble at their stop at Penang and had to string up the engine on a tree to repair it after it had seized up. They went on to make the longest non-stop flight between Timor and Darwin and arrived with a fuel tank which was literally empty.

The US Army, with a Fokker F III (T-2) transport machine also got into the long-range flying business. In 1923 two pilots, McCready and Kelly, made the first non-stop, coast-to-coast flight in USA from New York to San Diego.

While these trail blazers were showing what could be done, the aircraft manufacturers and the infant airlines were getting on with the more prosaic business of earning a living. Ex-service

Alcock and Brown set off from Newfoundland on their epic transatlantic flight by Vickers Vimy in 1919

machines were all very well, and quite admirable as far as they went. But they had not been built to carry passengers–for instance, the otherwise excellent Vickers Vimy had far too slender a fuselage for airline work, and those used in commercial service, by such firms as Instone Air Line, had a new, deep, 10-seat cabin fuselage. The service machines had been built regardless of cost and with little thought for economic operation, and most required quite specialised servicing. Snags like these revealed themselves as the months went by, and the time was ripe for new designs purpose built for carrying passengers.

The most significant step of all in the early 1920s was the introduction into service of all-metal aircraft, though the 'stick and string' type of machine with fabric covering was destined to remain in service–and in some cases in production–until 1940 and beyond. The significance of metal structures for aircraft was not, in fact, widely appreciated at the time. In rather conservative fashion, British airlines went on ordering and using fabric-covered machines long after all-metal aircraft were in service elsewhere, notably in France, Germany and America.

The first all-metal aircraft to appear specially for the airliner market was, in retrospect, years ahead of its time–the Junkers J.13 (which was later redesignated F.13). It was a low-wing, angular, single-engine monoplane with fully-enclosed cabin. Its skinning was corrugated for added strength–this became a characteristic of other Junkers designs–and the appearance was in complete contrast to the many fabric-covered biplanes then in evidence everywhere. The Junkers F.13 dispensed with struts and stays. The covering was duralumin and the wings were of cantilever type. There was a single 185 hp BMW engine and capacity for the pilot and four passengers. The Junkers F.13, it should be noted, was itself a development from the various all-metal warplanes–starting with the J.4–which the enterprising Junkers firm had built in the 1916–18 period.

The Junkers F13 was a four-passenger, all-metal monoplane. It was the workhorse of the Lufthansa Fleet between 1926 and 1932, and saw extensive service with other companies

Junkers J.13s went to USA as mail planes and were to be seen flying all over the world for the next 20 years or so, Lufthansa, the German national airline, being included among major users.

Britain (or at least the Short Bros. and Harland concern) was actually ahead of the Germans by 1920. The new Short aircraft was the Silver Streak which appeared in the 1920 Aero Show at Olympia. It was a biplane, but it proved superior to the Junkers technically, having duralumin stressed skin covering and a duralumin framework. But this advanced machine attracted little attention when it was first announced and the chance–seen in hindsight–to lead the world with this type of aircraft was lost.

The young Anthony Fokker returned to his native Holland after 1918–following his brilliant stint as a designer for the Germans in the war. He now began an equally fruitful period as a designer and builder of aircraft matched beautifully to the needs of the commercial market of the time. His high-wing designs had metal framing with plywood-covered wings and fabric elsewhere. The new Dutch airline KLM went to Fokker with their requirements and the F.II was the result, a high-wing monoplane with the same 185 hp BMW engine as the Junkers J.13. It was slower though.

Fokker's classic design for all time, however, was the very widely used F.VII, which came with many variations and modifications in 1926. The F.VII was a 10 seater and the mechanical arrangement was such that almost any radial engine of suitable power could be used. This was handy when the aircraft was built under licence in many countries of the world and a local engine could be used to suit the customer's requirements. A quick development was the F.VII B-3m, essentially the F.VII with two extra motors slung one under each wing. The most famous of these Fokker Trimotors was Sir Charles Kingsford-Smith's 'Southern Cross' which made record-breaking flights to Africa and Australia. Fokker did so well in Europe with his small airliners that he opened up a factory in New Jersey, USA. This encouraged an American designer, William B. Stout, to produce what amounted to an American version of the Trimotor, and in his turn, the motor car magnate Henry Ford bought out Stout's plant and design

above
A Fokker F.III, developed in Holland to the special requirements of the Dutch airline KLM

top
The Fokker F–10 Super Trimotor was built in the USA by Fokker's American offshoot

Ford Trimotor shown here being refuelled was still flying in regular airline serice in the 1960s with an US internal line. It was rumoured that a Fokker was surreptitiously measured by Ford designers before the 'Tin Goose' was finalised

and shipped everything to Detroit. Thus began the construction of the immortal Ford Trimotor which served American civil aviation well–so well that a few Trimotors were still to be encountered in commercial service in the 1970s while the basic design, in an updated form, was again in production in the 1960s. Unlike the Fokker, the Ford was an all-metal job, it was big and it was tough, so withstood immense punishment on indifferent landing grounds. The aircraft soon earned itself an unforgettable nickname, the "Tin Goose".

In Britain the major aircraft builders were still staying with conventional 'stick and string' machines, though they were largely designed to the order of the major commercial airlines. The newer aircraft were stately enough machines–the D.H.54 and D.H.66 were fine aircraft, and in the late 1920s came a big breakthrough with the very elegant Armstrong Whitworth Argosy. In this machine Imperial Airways had for the first time a cabin staff, a single steward, who served meals in flight on what they called their 'Silver Wing' service from London to Paris.

Such types as the de Havilland D.H.54, D.H.66 and Armstrong Whitworth Argosy were biplanes in the old tradition, fabric, struts, rigging, and open cockpits, in contrast to the increasing number of metal aircraft appearing elsewhere. In general, though, Imperial Airways, as implied by the name, concentrated on opening up routes across the British Empire rather than intensive European operations. This led to a new generation of big airliners in the early 1930s which were certainly grand and impressive in concept but were hardly any advance on their predecessors in technological terms. Best remembered of all, and by far the most imposing, were the luxurious Handley Page H.P.42 and H.P.45 biplanes. The former (24 passengers) was used on routes beyond Europe, while the latter (38 passengers) were used in Europe. There was an enclosed crew cockpit (the first in a British airliner). The aircraft, of which both types were externally similar, spanned 130 ft and were 90 ft long, with four Bristol Jupiter radial engines. The speed was about 100 mph. The Short Scylla and Armstrong Whitworth Atalanta–a high-wing monoplane–were other Imperial Airways types of the earlier 1930s.

By contrast, elsewhere in Europe, Fokker and Junkers were the most prolific of airliner producers. Junkers led the way with very advanced, all-metal aircraft designs. A fine, low-wing, tri-motor transport, the Junkers G.31, appeared in 1929, paving the way for the similar but bigger Junkers 52/3m of 1932, which became the mainstay of operations for Lufthansa, the German national airline. Indeed, the Junkers 52 was one of the most widely used military transports in the Second World War and remained in service with one or two operators until the 1960s, and a trio were in Swiss Air Force service well into the 1970s. With three 575 hp BMW radial engines the Ju 52 had a wingspan of 96 ft, was 62 ft long, carried 17 passengers and cruised at 152 mph. It is interesting to note that the lofty H.P.42 was an almost exact contemporary of the Junkers design. In the technology of all-metal aircraft, however, even the Junkers 52/3m was outmoded. There was a limitation to the effectiveness of the corrugated metal skin, with drag and low tensile strength being two major problems inherent in its use. Short, with their

below, left
Kingsford-Smith's Fokker Trimotor 'Southern Cross' at Croydon in 1930

below
A Fokker F.XII in Danish airline service

left
The Fokker F.XX with Wright Cyclone engines, which appeared in 1933, was the most refined of the trimotor designs, with, for example, a retractable undercarriage. It carried twelve passengers

Silver Streak of 1920, and Boeing with a superb, fast mailplane, the Monomail of 1930, had shown that smooth, stressed-skin construction was superior for aircraft. Essentially this consisted of light alloy covering panels, pre-shaped and riveted to an alloy-framing so that the skin carried some of the stresses and loads. The resulting airframe, resistant to twists and heavy stress, enabled such traditional aircraft features as struts and rigging wires to be dispensed with. After the Junkers 52, even that go-ahead company realised the superiority of stressed skin over corrugated skin and used this in their subsequent designs.

It was in the United States that some of the biggest strides were taken in the development of air transport, and the innovations soon took the country to pre-eminence in the civil aviation field. Curiously enough, however, the United States was the last of the major nations to establish any kind of passenger-carrying scheduled airlines. While commercial carriers were criss-crossing Europe in ever-increasing numbers from 1920 onwards, in USA there were only the mailplanes flying short-haul routes to open up a network of air communications across the country. The slow start to passenger-carrying airline activities is partly due to the limitation of range of early aircraft–distances between major cities tended to be greater in USA than in Europe–and the fact that both rail and road communication in the 1920s in America was fast and efficient. Until 1926 the mail services were operated by the US Post Office, which kept a monopoly of the route flying. In that year, however, the government took over administration of the routes and navigation aids and threw open the flying operations to private tender. This at last gave an impetus to passenger flying, which really started in April 1927 when Colonial Airlines started scheduled services between New York and Boston.

The very first American airliners were, in fact, mailplanes modified in design to carry passengers, typically four in number. One such classic type was the Stinson Detroiter which was put into service with North West Airlines in 1927. A neat biplane with single 220 hp radial engine, the Detroiter was in service as a light airliner well into the 1930s.

Most of the established American airlines, in fact, started life in this period as mail carriers. In essence each route (33 in 1928) was opened to tender and the lowest tender usually secured the job. The most spectacular route was the longest, Chicago to San Francisco, and this prize contract went to the Boeing aircraft company. The route started operations on 1 July 1927, the first full transcontinental regular service. It was a huge success for Boeing who had designed a passenger-carrying aircraft specially for the job. This was the Boeing 40A, a biplane with the pilot seated well back in the fuselage and two passengers and mail in an enclosed cabin right forward. The success of the Boeing 40A in operating this long route efficiently—against all predictions of failure—was largely due to the single Pratt and Whitney Wasp radial engine of 420 hp, which first interested Boeing for installation in warplanes but was soon to become a widely-used motor in air transport too. To operate the Chicago to San Francisco route, Boeing formed their own line, Boeing Air Transport. The route prospered mightily, and by 1930 was being operated by a tri-motor biplane, the Boeing 80A, with three Pratt and Whitney Hornet 525 hp engines, carrying 12 passengers in a fully enclosed cabin. The cruising speed was about 115 mph. A big innovation of the time was the employment of air stewardesses, the first ever, on this major route.

All the early running was now being made by Boeing, at least in so far as significant developments were concerned. The all-metal monoplane, the Boeing Monomail, appeared in 1930, way ahead of any other machine in concept and performance. It had a single 575 hp Hornet engine, stressed-skin construction and a retracting undercarriage. From this was developed an all-metal twin-engined bomber, the Boeing B-9, which appeared in late 1931, and was a low winged monoplane, with many of the advanced characteristics of the Monomail. While the B-9 was not accepted for production by the US Army it did, in effect, act as the prototype for Boeing's 'breakthrough' design, the 247 airliner which totally revolutionised the whole concept of air passenger work.

Because of the experience with the B-9, Boeing Air Transport

below, left
The Junkers Ju 52 entered Lufthansa service in 1933 and was to become one of the most famous machines in aviation history

below
A Curtiss Condor of American Airlines, 1934. It was the last biplane airliner in service in America

were able to order the 247 'off the drawing board'. Some 60 machines were put into production, the first flying in February 1933. Among the features of the revolutionary Boeing 247 were all-metal, stressed-skin construction, ability to fly (and climb) with one engine, fully enclosed crew and passenger compartments, rubber de-icing 'boots', variable-pitch propellers (in the second production model), trim-tabs, streamlined faired engine nacelles integral with wings, automatic pilot, and retractable undercarriage. In short, the Boeing 247 was the first 'modern' airliner, with a layout that was followed by nearly all subsequent propeller-driven airliners. It revolutionised airliner design and operation; a true milestone in the progress of aviation development. (It is an interesting point, too, that Boeing's ability to utilise common design features for both military and commercial projects with consequent huge savings in time and production has not been lost on the company since–most of Boeings many successful airliner designs since the 247 have had a successful military counterpart.)

The Boeing 247 in service (a later model with refinements was the 247D) carried 10 passengers and 400 lb of mail. The Chicago-San Francisco flight time was taken below 20 hours. The span of the aircraft was 74 ft, length $51\frac{1}{3}$ ft, top speed 182 mph (247) and 200 mph (247D), and cruise speed was 155 mph (247) and 189 mph (247D). Two machines were supplied to the German airline Lufthansa, but the rest went to Boeing Air Transport, or more accurately United Air Lines, a company formed in 1934 by merging Boeing Air Transport with two other lines to make America's largest internal air carrier.

The appearance of the magnificent Boeing 247 highlighted the massive technological advances which made such aircraft possible by the mid 1930s. We have already noted the change from fabric to metal covering which was fundamental for strength. Engine development went hand in hand with this structural change. The Wright Aeronautical Corp. in USA produced a very powerful radial engine, the Whirlwind, of 220 hp. Some breakaway designers from Wright produced an even more powerful engine, the Pratt and Whitney Wasp of 400 hp which became one of the most widely used of all aero engines. Indeed, these powerful radial engines with nine cylinders, which eliminated the complications found in liquid-cooled, in-line engines, made bigger aircraft with bigger payloads and longer ranges possible. It was estimated, for example, that the weight saved over an in-line, liquid-cooled engine by the Whirlwind radial allowed an aircraft to carry fuel for an extra 200-miles range. These radial engines were pleasantly reliable, too, and easy to service and change. Not surprisingly they were either licence-built or closely copied by several other nations including Russia and Japan. The Wright and Pratt and Whitney companies went on to build whole families of radial engines developed from their original designs, and there were still many aircraft flying with this type of engine years later, even in the 1970s. In parallel with the new engines went the variable pitch airscrew. The blades were set to a fine pitch or angle for best possible power output at take-off, then pivoted (through gearing in the propeller hub) to a much coarser or shallower angle for cruising flight when the engine could be throttled down. The Boeing 247D introduced this feature on airliners, though it was concurrently appearing on some fighter planes.

The DC–3 remained in service forty years after it first flew. This machine served with the British independent line Air Anglia in the 1970s

The monoplane configuration became firmly established with all-metal aircraft. Both high- and low-wing placings were common in new designs, but the latter was favoured in the advanced designs which featured retracting undercarriages and streamlining. Biplanes stayed in vogue with some nations and designers longer than others, but the low-wing monoplane was soon established as essential for high-speed aircraft, America giving the lead in this respect.

Another big technological advance was in the field of flight instrumentation. In the 1920s instruments were still very basic, at best a speed indicator and a magnetic compass. Once such developments as the US mail routes with proper schedules were fixed, there was the need for suitable devices to make such trips possible whatever the weather and conditions. The gyroscope was adapted by Elmer Sperry who had worked on the problem since 1912. This resulted in Sperry's autopilot device (1932) and the gyro compass, plus many other related instruments such as the earth inductor compass, the artificial horizon, turn and bank indicator and an accurate type of altimeter. The Guggenheim Foundation in 1928 sponsored research into 'blind' flying, and in September 1929 the results of the work were indicated when Lieutenant James Doolittle made the first American take-off and landing in thick fog and with his cockpit shrouded. Similar experiments were being carried out in Britain at about the same time. Radio beacon beams, in conjunction with the new generation of instruments, made this technique of blind flying a perfectly practical proposition. Aircraft such as the Boeing 247 and later types incorporated all these improvements.

Trim tabs were also introduced on the Boeing 247–now a commonplace idea–which enabled slight changes to the setting of the flying control surfaces without major changes in the control positions. A secondary change in aircraft construction in the late 1920s, as an alternative to metal skinning, was the bonded plywood surfacing replacing fabric. The Lockheed Vega of 1927 and Lockheed Air Express mailplane were among many small aircraft types in which plywood skinning was an extensive feature.

While the Boeing 247 was the first of the modern concept airliners to enter service, it was certainly not alone in competition for the airliner market, and there were others following close behind. In 1930 another major American airline had been formed, TWA–now Trans-World Airlines, then

Transcontinental and Western Air, by the merger of four smaller lines. The major types operated by the new company were the Fokker and Ford Trimotors, plus an assortment of other, smaller types. To rationalise the fleet, TWA asked the Douglas company to build a new tri-motor design. An engineer, Arthur Raymond, was given the job, but was unimpressed with the vibration and noise caused by the tri-motor configuration. He decided to eliminate the fuselage engine by using more powerful Wright engines in the wing, two 710 hp Cyclones. Using all the latest known techniques—stressed-metal skin, split flaps, retractable undercarriage with brakes and full hydraulics, soundproofed interior and luxuries like reading lights—Raymond came up with the DC-1 prototype. It first flew in July 1933, some five months after the Boeing 247. It exceeded TWA's requirements, moreover, carrying 14 passengers at 178 mph, rather than 12 at 145 mph. The DC-1 was quickly followed by the DC-2, virtually the production version of the DC-1. TWA ordered 20 straight away, doubling the order soon afterwards. The DC-2 rapidly eclipsed the Boeing 247, being clearly a superior design with much improved passenger comfort and the more impressive performance. About 200 were built and a number were sold to European airlines (Swissair and KLM were two early operators). The DC-2 made a big impact on the world when a KLM machine (and a Boeing 247D) was entered in the London-Melbourne air race of October 1934. Specially-prepared aircraft were built and entered for this race, but the two airliners were almost standard *and* carried a payload. They came second and third respectively, beaten only by the de Havilland D.H.88 Comet, a specially-built racing machine. The aviation world was stunned by the achievement.

Douglas did not rest on their laurels—better was to come. An enlarged design, similar in layout but bigger, was produced and put into production. This was the DC-3, destined to become the most legendary civil aircraft of all time; some would say the most legendary aircraft of all and leave it at that! For the DC-3 simply overshadowed the competition for years afterwards.

The DC-3, however, came about to meet the requirements of yet another internal US carrier, American Airlines. They were operating the last big biplane airliner type in America, the Curtiss Condor, on a popular coast-to-coast sleeper service. The airline had to compete, however, with the lines offering Boeing 247s and DC-2s, neither of which were wide enough to fit sleeping berths. Douglas simply widened the fuselage to take 14 sleeping berths; the type was initially called the DST—Douglas Sleeper Transport—but it was then offered as a 21-seat, standard airliner in its better known form as the DC-3. Coast to coast, American Airlines operated a 17 hour, 45 minute westbound schedule and 16 hour eastbound schedule. Two 1,000 hp Wright Cyclone engines powered the early machines, but 1,200 hp Pratt and Whitney Twin Wasps later became standard. The DC-3 had a wing span of 95 ft (85 ft in the DC-2) and a length of $64\frac{1}{2}$ ft (62 ft in the DC-2). The loaded weight was 25,200 lb, cruise speed about 180 mph, and the range 1,500 miles (1,193 miles in the DC-2). The range alone shows how effectively the Douglas aircraft upstaged the Boeing 247 which had a range of only 485 miles.

A Lockheed Orion in Swissair service in 1930

By December 1941 over 800 DC-3s had been built and it was adopted as the standard Allied transport aircraft in the Second World War, when some 10,123 further examples were built, under various designations; it was best known as the C-47 Dakota. Post-war, it was in service in prodigious numbers and in addition had been licence-built in Russia as the PS-84 (later the Li-2). In 1976, 40 years after it first flew, there were still hundreds in service. The DC-3 took massive punishment and saw rigorous service that would have destroyed lesser planes. Its longevity was something of an accident. When it was designed no great attention was given to the problem of metal fatigue, but the multi-cellular structure designed into the DC-3 wings was quite fortuitously tough. Some DC-3s had a fatigue life of over 60,000 hours compared with around 10,000 hours for lesser aircraft. So the DC-3 could fly on long after its contemporaries had been consigned to the scrapyard.

There was one remaining star in the firmament of dynamic American aircraft companies which made an impact on the airliner scene in the 1930s—the Lockheed Aircraft Corporation. Lockheed had started with some successful small aircraft. One such was the Lockheed Orion which first flew in 1930. A single-engine feeder liner, it was the very first with a fully retracting undercarriage. Swissair were major European users of this machine. The Orion was followed by the twin-engined Electra in 1934.

The Electra had two 420 hp Pratt and Whitney Wasp Juniors. It followed the layout of the Boeing 247, but had twin tails. It carried 10 passengers and had a top speed of 203 mph, with a cruise speed of 180 mph; span was 55 ft, length $38\frac{1}{2}$ ft, and weight 10,500 lb. Northwest Airlines were the first users of the Electra while Braniff and Pan American were among other operators of the 148 which were built. The aircraft sold well to airlines all over the world. The Electra was notable in being one of the first American airliners ordered by Britain—British Airways, formed for European operations in 1935, had seven of them and one flew Prime Minister Chamberlain to Munich for his famous 'peace in our time' meeting with Hitler in October 1938.

Two of the early all-metal
airliners of the late 1930s—the
Beech 18 *(left)* and Lockheed
Electra—still in service in
Australia in the early 1970s

An improved derivative of 1936 was the Lockheed 12, a slightly smaller machine with Wasp Junior engines. The most successful of the series, however, was the handsome Lockheed 14, which appeared in 1937 and introduced a new feature, the Fowler flap, which could be extended from the trailing edge of the wings for take-off. On landing it was used to increase drag and reduce approach speed. Constant speed propellers which governed the pitch according to engine speed were another advanced feature first introduced to airliners on the Lockheed 14. This technically-advanced machine which cruised at 190 mph and carried 14 passengers in comfort was sold extensively in the late 1930s. And with the addition of a turret and armament it was to see much war service as the Lockheed Hudson in the Second World War.

Faced with such rapid technical development in airliners in America, some of the European constructors began to produce designs of a comparable type. The Fiat G.18 which appeared in 1935 was a twin-engine machine with a range, performance, and dimensions close to the DC-2. Only nine machines of two types (G.18 and G.18V) were built, however, all for the Italian national airline Alitalia. The French contemporary was the Dewoitine 332, a handsome tri-motor. The prototype of 1933 crashed, but its derivative, the D.333, and a further development, the D.338, were successful machines which served Air France, some of the latter still being in use after 1945. The Bloch 220 was a 'new generation' design owing something to American developments. On the London-Paris route it took just over the hour for the journey. It was a low-wing monoplane carrying 16 passengers. Germany made most of the running in Europe, with the 10-passenger, fast, twin-engine Heinkel 111, the single-engine Heinkel 70 feeder liner, and the Junkers 86 as the most notable types with Lufthansa from the mid 1930s.

By the late 1930s, airlines and airliners had come a very long way in well under 20 years. From stick, string and fabric construction of the most primitive sort to all-metal monoplanes of advanced design was an enormous jump. And in 1938 the airlines were poised for yet another step forward, trans-oceanic flight which would really link up the continents of the world in a spectacular way.

The Light Aircraft Era

While the airlines grew up between the two world wars, another sort of civil flying was evolving along its own distinct lines. This was 'flying for everyman', a far-sighted dream in the 1920s when heady enthusiasts–flyers and constructors–foresaw the day when small aircraft would be used for personal transport rather like the motor car. Some even went so far as to prophesy the day when light aircraft would completely supplant cars and when the individual with his own aircraft would be free to fly anywhere–or at least anywhere his aircraft could take him. Those early dreams have never been realised, though the United States has come closer to the vision than any other nation. Another war intervened just as the light aircraft was coming of age, and economic and physical conditions since have prevented any full realisation of those earlier prophesies.

Even before 1914 the pioneers were offering production 'light' aircraft to any buyer with the money. Santos-Dumont designed his Demoiselle as a light personal aircraft, as we have already seen, while Blériot, Voisin, the Farmans, and Avro were among early constructors who offered machines for sale. There were some sales, too, to the affluent and daring, but the 'flying for everyman' idea was still far off.

D.H.51

The Avro 504N, developed from the 504K, was a popular civil type for 'joy riding'

The First World War gave a big impetus to the development of light aircraft, just as it also set airlines on the way to growth and ultimate success. The reasons were similar; thousands of young men had learned to fly and many of them were keen to utilise their new skills on return to civilian life. Some were so keen on flying that they became professional pilots, but many more were happy to take up flying simply as a pastime. There were spare aircraft in abundance, too, thousands of redundant fighter and trainer machines, all single-engine types, being sold as war surplus for nominal sums; $300 or so would secure a Curtiss Jenny. With a few notable exceptions, though, surplus warplanes were not the best or most successful for private flying. A fighter, built with no great regard for costs, was extremely expensive to maintain and operate. The fighters that were sold mainly went to commercial operators. Skywriting–putting out smoke-written advertising slogans in the sky–or banner-towing were the main employ of ex-fighter types. Indeed, one reason why some First World War fighters were ultimately preserved as museum pieces was because they were skywriting machines which survived a decade or more after 1918, so that when they were finally pensioned off their historic value was realised.

The aircraft which found a real niche in the post-war flying scene were the trainers. Slower, more economical, generally tougher and simpler than the fighters, they were forgiving and durable. In Britain the most familiar and famous of all was the ex-RFC trainer, the Avro 504K. The 504 was a direct link with pre-war days, for its prototype flew in 1912. It was a beautifully responsive machine, fully aerobatic and light to handle, and powered by the Gnome rotary engine (though some models had other engines). The 504's American equivalent was the Curtiss JN-4 Jenny, which was built (and sold surplus) in vast numbers. It was the standard trainer designed for the US Army Air Corps, and like the Avro 504K it was almost viceless. Most had a 90 hp V8 engine, but some had a 150 hp Hispano unit. The D.H.4, licence-built in USA was another popular machine for civil use.

On both sides of the Atlantic, but most brashly in America, the early 1920s was the era of the 'Flying Circus' and the 'Barnstormer'. Young men with their Curtiss Jennys would get together or fly alone, would follow a travelling fair or circus, or simply hire a field from a local farmer. Here they would stunt, race, and give joy rides. The stunts really pulled in the customers. Wing walking, tied-together flights, aerobatics, girl acrobats harnessed below or above the wings, dummy bombing raids, and parachute jumping were guaranteed to thrill a public whose previous knowledge of aviation was largely gathered from the popular press or the cinema pictures. The show usually ended with some joy riding–a trip over and around the adjacent countryside for such sums as 5s (50 cents) to £4 ($8) a head depending on the ambience, or the size of the circus deciding the fare level. In Britain a leading 'Air Circus' operator was Alan (later Sir Alan) Cobham who went on to rather greater things in the service of British aviation. In truth the Air Circus and Barnstorming ideas did little to further the progress of the light aircraft in the technical sense, but they did produce a breed of skilful, daring pilots, and they made people air-minded. Many an embryo pilot of the next generation became distinctly enthusiastic as a result of a joy ride at an Air Circus. Conversely, the pilots could make a living of sorts–sometimes a good one–in a period of general commercial decline.

Meanwhile, as the First World War progressed some serious thought was being given to the 'flying for everyman' concept. In Britain a tiny machine with big potential was the Grain Kitten which first flew in 1918. It saw no production. But the Avro concern was also active and in 1919 they produced a prototype specifically designed for the light plane market, the Avro Baby. This was a diminutive single seater with a 35 hp engine. A second model was a two-seater with a 60 hp Cirrus engine. The Baby got a lot of publicity when Avro's test pilot, Bert Hinkler, used it for some of his famous long-distance flights. Avro themselves, plus other makers, tried to improve on the basic light plane idea, though some of the prototypes were never put into production. Outstanding was the Parnall Pixie of 1923, which was a monoplane offered in single- and two-seater forms. With a 32 hp Bristol Cherub engine, it had folding wings to assist in stowage. The Beardmore Wee-ke, Hawker Cygnet, and Avro 506 were all similar types, the Cygnet a biplane, the rest monoplanes. Smallest of all was the de Havilland D.H.53.

These various machines were all in the 'ultra light' class and the design of most was spurred on in Britain by the Light Aeroplane Competitions of 1923 and 1924. These were sponsored by the government and supported by the *Daily Mail* newspaper and leading industrialists, with the avowed aim of stimulating the development of light, economical designs for owner-pilots. The promising designs, like the Pixie, were quite shortlived, however. While 'official' opinion was that the average owner-pilot would be satisfied with a lightweight machine having a simple airframe and maximum economy, low-power engine, there was another school of thought which considered that the modern pilot would want more luxury, passenger and luggage space, and ample power and range to give a good reserve capability for emergencies or longer flights.

One designer who drew his own conclusions about this was Geoffrey de Havilland, whose own contribution to the Light

Light aircraft flying in 1929. Sir Henry Seagrave about to fly his D.H. Moth from Lympne

top
D.H.60 Moth

above
D.H.82A Tiger Moth

Aeroplane Trials of 1923 was the D.H.53, a dainty little monoplane with a 6 hp motor-cycle engine. Though pleased with the D.H.53, de Havilland thought he could do better. He wasted no time entering the 1924 trials and instead designed the type of machine he would personally find desirable. The result was a 30 ft wingspan biplane (this dimension matched the D.H.53). It weighed about 770 lb (twice the D.H.53's weight), and it carried a name which was to become historic and immortal–the de Havilland D.H.60 Moth. The well-proven layout of the First World War fighter type was followed (de Havilland had been an active designer in the war), so the cockpits were in tandem, the wings were single bay with a facility for folding for hangar or garage storage, and the structure, part metal, part wood, followed conventional box-girder style. There was a 60 hp Cirrus in-line engine giving a top speed of about 90 mph. The cost was £595 and the prototype flew in February 1925. It was immensely successful and sold well. The British government had been persuaded to allow flying clubs to be set up aided by subsidy, to encourage light aircraft sales and production. The Moth did well out of this opportunity and the clubs and members alike were good customers for the new machine. Competitors appeared as well, the Avro Avian (a cheaper machine) making the biggest impact, but there was also the Aero Club Cadet, the Blackburn Bluebird, the Spartan Arrow and a parasol-wing monoplane, the Westland Widgeon III, which enjoyed some success.

De Havilland's Moth dominated the British and Commonwealth market, however, and the D.H.60 started off a notable dynasty of de Havilland light aircraft in the late 1920s and 1930s in what became the only 'golden age' for light aircraft in Britain. There were improved variants of the Moth including the Gipsy Moth of 1928 with a 98 hp Gipsy engine and a D.H.60M in 1929 with a metal-covered instead of fabric fuselage. There followed a D.H.60T trainer, which was adopted by the Royal Air Force, and from this was developed the D.H.82 Tiger Moth, for both RAF and civilian service. Appearing in 1931, the Tiger Moth was built in thousands up to 1945, having been adopted as a standard service trainer. Many were still flying in the 1970s, a remarkable tribute to Geoffrey de Havilland's original perceptive design. Variations on the theme were many, among them the D.H.80 Puss Moth of 1930, a high-wing cabin monoplane, the D.H.83 Fox Moth of 1932, a biplane with forward passenger compartment like the American mailplanes, the D.H.85 Leopard Moth of 1933 (an enlarged version of the Puss Moth), and the D.H.87 Hornet Moth, a biplane cabin aircraft of 1934. De Havilland also built a monoplane racing version of the original Moth in 1927, called the D.H.71 Tiger Moth, with a top speed of 187 mph. In 1938 they revived the monoplane idea with the D.H.94 Moth Minor trainer which was supplied to the Australian government as a trainer but was also sold in Europe. Finally, mention must be made in passing of the series of graceful biplane feeder liners built by de Havilland in the 1930s. The 6-8 passenger D.H.84 Dragon of 1932 and its successor, the D.H.89 Dragon Rapide of 1934 were the best known, the latter seeing RAF war service too with some surviving well into the 1960s in airline service. Less successful was the 4-5 seat D.H.90 Dragonfly of 1935 which was an attempt to provide a twin-engine, multi-seat personal aircraft to complete

D.H.89 Rapide still in British European Airways service in the 1960s as an Islander

the de Havilland 'family' of light aircraft. It found few purchasers, however, among private pilots and was mostly used as an ultra-small feeder liner or charter machine—an early 'air taxi'.

Another big name in the British light aircraft scene in the 1930s was that of Miles—which encompassed two brothers, F. G. Miles and George Miles, who grew up with the age of aviation. F. G. had a joy ride at the age of 22 just after the First World War. This made him decide to go into aviation and he produced a design called the Gnat, with a 650 cc motor-cycle engine for the 1923 Light Aeroplane Trials. Realising that he could not fly, Miles went to Cecil Pashley, an instructor, who was a veteran of the pre-war Graham-White era, and proposed some lessons, plus the idea of setting up a firm to give joy rides and to design aircraft. Pashley owned an ex-RAF Avro 504K and with this and some assorted types acquired elsewhere they started giving flying displays and joy rides from a field they rented at Shoreham in Sussex (near to a disused pre-war flying field). Among aircraft purchased was an early small light plane of 1920, the Avro Baby. Miles had the idea of modernising it and in 1929, with new, powerful engine and an enlarged tail, it became the first Miles aircraft, the Southern Martlet. Nine were built. In 1932 came the Miles Hawk, a new, clean, monoplane design to compete with the Moth and Avian biplanes. It had a 95 hp Cirrus engine and cruised at over 100 mph; about 55 were built and the firm was by now fully established as a constructor.

A whole range of derivatives appeared through the 1930s

above
D.H.83 Fox Moth

opposite, top
A pair of French Rallye light aircraft, the 100S Sport (foreground) and a 180GT—the designations appear to owe something to the motor industry!

opposite, lower
Another recent French light aircraft is the Robin

below
Auster Aiglet of 1950

including the Hawk Major, the Hawk Trainer, and Hawk Speed Six, all basically similar tandem-cockpit layout monoplanes. The Speed Six, however, was a single-seat racing version with 200 hp Gipsy engine and a top speed of 195 mph. The Hawk Trainer got itself a good reputation in flying schools as a safe, docile, stable machine, and when the RAF was expanded from 1936 a developed version of the Hawk Trainer, named the Magister, was selected for service use and over 1,200 were built. This contemporary of the Tiger Moth survived as a civil flying machine into the 1960s in small numbers. Miles also built cabin monoplanes, the Falcon Major (1934), Falcon Six (1935), the very successful Whitney Straight (1936), Monarch (1938) and M.28 (1939) leading directly to the Messenger of 1942. This was built originally as an air observation 'spotter' plane for the British Army and in post-war years was popular in what became a much diminished, light aircraft market. In 1946 a twin-engined version of the Messenger was built, a highly successful type called the Gemini which had two 100 hp Cirrus engines, a 36 ft span, and a top speed of 146 mph.

Not strictly speaking in the light aircraft field, but arising from the same activities, was the most celebrated of the Miles designs, the fast Miles Master training aircraft of the RAF. Produced as a 'private venture' by Miles, it proved to be just what the RAF needed to replace biplane advanced trainers in order to familiarise pilots with monoplane characteristics for the new fighters of 1939–40. About 3,000 Masters in three versions were built by 1942.

Edgar Percival was the other great luminary of British light aviation in the 1930s and his achievements were in many ways to parallel those of de Havilland and Miles. The Percival Gull of 1932 was a fast (155 mph) three-seat cabin monoplane more advanced in style and concept than its contemporaries. In 1934 a more powerful version, the Gull Six, appeared, having a streamlined spatted undercarriage, large cabin, and 200 hp Gipsy Six engine. From this was developed the Vega Gull of 1936, externally similar, and this in turn led to the Proctor of 1938 which was adopted for service use by the RAF and Royal Navy as a general utility light plane. Over 1,000 were built during the Second World War period, and 'demobilised' Proctors saw extensive use on the British light aviation scene until well into the 1960s. The most attractive Percival aircraft was the Mew Gull racer of 1936, a sleek single seater with a top speed of 230 mph. It was prominent and successful in the air racing scene until 1939.

All this intensive activity in British light aviation was popularised by the deeds and achievements of numerous adventurous young pilots who became the folk heroes of the

The Percival Gull 4 of 1933

Beech Duke

day. Among them was Amy Johnson, a shorthand typist who learned to fly in a Moth in 1928, and in 1930 became the first woman to fly from England to Australia in her dark green Moth 'Jason'. It was an epic of endurance which hit the headlines, and even inspired a popular song, 'Wonderful Amy'! Amy went on to make more great solo flights, and subsequently married another famous solo flyer of the day, Jim Mollinson, who had also flown a Moth to Australia (in 1931) and, more spectacularly, made the first Atlantic crossing from east to west by a lone pilot, flying a Puss Moth in August 1932. The pioneer solo flyer to Australia had been another popular hero, Bert Hinkler, in an Avro Avian in 1928, an achievement which inspired Amy Johnson to be the first woman to make the same flight. A later heroine of the air was Jean Batten, who made a record-breaking solo flight from England to New Zealand in 1936, piloting a Percival Gull. Air races became popular spectator sports, the Kings Cup series being one of the premier events in the world calendar. Empire Air Day at Hendon, though basically a service flying display, was highly spectacular and put over the excitement of the aviation age to thousands of eager spectators who flocked to Hendon for the event. For Britain and the Empire, the fact that the new air age caught the imagination of thousands of young men was not a bad thing. Many learned to fly at the local clubs, others took more than a passing interest–flying model aircraft also became a popular hobby in

The Pou-du-Ciel or 'Flying Flea' midget aircraft

the 1930s – and when a scheme was started by the Air Ministry in 1936 for local Reserve Flying Schools, many more took the opportunity to learn to fly coupled with the undertaking to serve in the RAF if war was declared. The result of all this interest and activity in light aviation paid good dividends in 1939–40, for Britain could immediately call upon a well-trained nucleus of pilots for an air force which was being rapidly expanded. A good proportion of the 'famous few' of the Battle of Britain were young men who had learned to fly at their local flying clubs a few years before the war.

In other European countries there was activity of a similar kind. Belgium contributed a classic light aircraft, the Stampe SV.4, which was an exact contemporary of Britain's Tiger Moth and very similar in appearance and performance – though considered the peer in aerobatic ability. With either a 125 hp Gipsy Major or 140 hp Renault 4P engine, it served as a trainer with the French and Belgian air forces as well as a favoured civilian type. After 1945 the Stampe went into production again and became particularly famous as a star aerobatic performer at post-war airshows. Many were still flying in Britain, France and Belgium in the 1970s.

From France came one of the most eccentric designs in light aviation, the Flying Flea, or Pou-du-Ciel (sky louse), designed in 1933 by Henri Mignet who had previously produced a more conventional light aircraft. Mignet believed in going back to pure basics, cutting out the complications, and producing an aircraft which anyone could make himself. It was a return to the philosophy behind the British 1923/24 Light Aeroplane Trials. It proved to be ingenious and practical, but a little hazardous. The Flea had 'tandem' wings, with the front 13 ft wing arranged to tilt for lateral control, and the incidence changed for longitudinal control. There were no separate ailerons or elevators. The pilot sat forward of the wings and there was a tiny motor-cycle engine in the nose. A British pilot saw Mignet's prototype and had one built himself. The Flea's tiny size and sheer impudence as an aircraft caught the public imagination. By 1936 over 100 had been built in Britain alone but a rapid succession of crashes in an 18-month period caused the aircraft's certificate of airworthiness to be withdrawn. Nonetheless, the Flea was not the only 'ultra-light' aircraft design to appear in the period 1925–39. There were dozens of them in America, Britain, France and elsewhere but most were one-off prototypes or very limited production machines which never made the impact of the bigger 'conventional' light aircraft like the Moth and the Gull. The Luton Buzzard (GB), the Drone (GB), the Aircamper (USA), Heath Parasol (USA), Lippish Storch (Germany), and Comper Swift (GB), were among the many ultra-light machines which were unveiled to the public. Of these, the Heath Parasol and Comper Swift saw considerable sales and success, but they were perfectly conventional aircraft with economical engines and scaled down to minimum dimensions.

The French government copied the British idea of subsidising flying clubs in 1931, in an 'Aviation Populaire' scheme, and this led inevitably to some keen light aircraft activity. Of many French types the Caudrons stand out. The Caudron Luciole was a 1928 design similar in style and size to the D.H. Moth, and as successful in France as the Moth was in Britain. Various engines were fitted and top speed was about 106 mph. The

Caudron Phalene was a high-wing cabin monoplane like the Puss Moth, and a later development was the Pelican. In 1935 came the Caudron Aiglon monoplane, not unlike the Miles Hawk Trainer in size and appearance. Salmson produced the Cri-Cri in 1936, an ugly but effective high-wing monoplane, and the Phrygane, a three-seat cabin monoplane. Morane-Saulnier were among the most prolific constructors, with some neat parasol-wing machines–the M.S.1811 of 1931, intended specially for 'Aviation Populaire' flying clubs or schools, the heftier M.S.230 (of which over 1,000 were built), the M.S.315, and the M.S.340 (of 1933). The Potez firm was also a maker of popular types, with the Potez 36 high-wing cabin monoplane of 1929, the Potez 43 of 1932, an improved model, and the Potez 60, a parasol-wing, 90 mph two seater built in large numbers for the subsidised flying clubs.

In the United States of America private flying on a big scale was developed a few years behind Great Britain, but once the US light aircraft industry was firmly established it started to dominate the world in this field. There was the major economic advantage of a vast home market allowing huge production runs for the most successful designs, plus the addition of technological developments–all-metal, stressed-skin construction and sophisticated instrumentation when required–such as were also incorporated in the new generation of American airliners. The pattern of development differed somewhat from that in Europe.

In 1924 there was a Light Plane Contest similar to those held in Britain. Henry Ford produced his Flivver at this time, an ultra-light 'Model T' of the air with a 20 hp, flat-twin engine. Assorted one-off, ultra-light types with names like the 'Bathtub' and 'Midget' appeared at this contest. These machines made

Lindbergh's celebrated 'Spirit of St Louis' New York–Paris aircraft

top
Postwar light aircraft from
America—the Cessna Skymaster

right
The Piper Super Cub of 1950

opposite, top
Piper Cherokee 6 of 1976

opposite, lower
Aeronca Sedan of 1950

above, left
The American light aircraft flying
scene in the 1930s. A Stinson
Reliant at a local airstrip at
Daytona Beach in 1935

above, right
An Aeronca C–3

little impact, however, and it was probably Colonel Charles Lindbergh's memorable New York-Paris solo flight of May 1927, in quest of a $25,000 prize that opened the public's eyes to the potential of the light aircraft. Lindbergh's machine 'Spirit of St. Louis' was a specially-adapted version of the Ryan B-1 Brougham, America's first light monoplane to be offered on the market as a production aircraft in 1926. The Brougham had a 200 hp Wright radial engine and was of high-wing configuration with a 4-5 seat cabin. Subsequently the Ryan company was sold up in the depression of 1929, but Claude Ryan started a new company in 1933 which was responsible for some fine aircraft, notably the Ryan ST-A which became a standard air corps trainer in 1939–45 as the PT-22.

Ryan's nearest competitor in the 1920s was the Stinson company, founded by the Stinson *sisters* in 1912 originally for exhibition flying and instructing, the two sisters, Kate and Madge, being the flying instructors. Their younger brother Eddie learned to fly during the First World War and afterwards took up instructing and had a spell of 'barnstorming'. Eddie had ideas for building his own light aircraft and realised this ambition in 1925 when he got some financial backing from Detroit businessmen. His neat, high-wing cabin monoplane, the Stinson SM-1 was very refined for its day, having an enclosed cabin, wheel brakes, and a self-starter. It had a 220 hp Wright radial engine, and hit the headlines with a near 13,000-mile solo flight from USA to Japan by a test pilot, Bill Brock, in a machine named the 'Pride of Detroit'. An enlarged version of the SM-1 was the famous Detroiter mailplane, but Stinson went on to build a classic series of light planes culminating in the Stinson Reliant, which saw war service in big numbers with US and British forces, and the Sentinal army 'spotter' plane and its civilian derivatives, the Voyager and Station Wagon, well over 2,000 of which were built. The early model of the Reliant (1935) was the first production light aircraft to have landing flaps as standard. In 1948 Stinsons were absorbed by another rival, Piper.

Piper was actually a latecomer in American light aviation, at least as a separate company. William T. Piper worked with Taylor Brothers Corp. who produced in 1931 the Taylorcraft Model A Cub. This was one of the significant designs which inspired the style and layout for a number of imitators. Various models were produced, typically with a 37 to 65 hp Continental engine. The Model A was a tandem seat, high-wing cabin monoplane of compact dimensions, span just over 35 ft, length

22½ ft. Cruising speed was about 74 mph and it was economical to run. In 1936 Piper bought out his fellow Taylor Brothers directors and put a modified version of the Cub design–called at first the 'New Cub'–into production as the Piper J-2 and J-3. It became the best seller of all time in the light aircraft field. Over 10,000 had been sold by the time America was swept into the world war in December 1941. The Piper Cub offered the market exactly what it wanted, when it wanted it–a small, personal aircraft of unfailing reliability which was cheap to buy and cheap to run. The Cub was produced as a military liaison and 'spotter' aircraft during the Second World War, resumed civil production in 1945 and was developed into the more powerful Super Cub in 1950. Over 20,000 Cubs of various types were produced and more than 200,000 pilots, it is estimated, must have learnt to fly in them. In the early 1970s there were still around 3,000 Cubs flying in the USA, and the second-hand price was often more than twice the original!

The Piper company went on to develop the theme with more success. Among their designs were the Cruiser three-seater of 1940, and Pacer four-seater of 1950 (with a later Tri-pacer offering

top
Piper Tri-pacer and the similar Caribbean (shown here) were direct developments from the Cub

above
The C–37 was an important Cessna type of 1937

Cessna Skylane

a tricycle undercarriage). Today Piper remains one of the 'Big Three' light aircraft producers in America.

Meanwhile, in 1937, the Taylor company was reformed with their own designs developed from the original Model A Cub. The Model B of 1939 was similar to the Cub but had side-by-side seating and several thousand were built before and after the war. The Model D of 1940 returned to tandem seating as a trainer aircraft and was selected in 1941 for use in the newly-established war emergency civil flying schools, and also for the Civil Air Patrol. A company in Britain was set up in 1940 to build the Model D under licence. With a 90 hp Cirrus engine replacing the Lycoming engine of the American version, the Plus Model D (as it was known) was ordered by the British Air Ministry as a 'spotter' plane with the name of Auster. This led to several wholly British developments of the original Taylorcraft design, and in 1946 the Taylorcraft firm in Britain changed its name to Auster Aircraft Ltd. and became the most successful of British post-war light aircraft builders with a host of further models in the next two decades. Thus the basic Cub design of the early 1930s influenced a whole generation of light aircraft on each side of the Atlantic.

Monoplanes were almost invariably the rule in the American market, but there was a notable exception in another classic machine, the Beechcraft 17R of 1932. Popularly known as the 'Staggerwing', it had the upper plane set behind the lower plane to give maximum visibility for the pilot. The 420 hp Wright engine gave a high speed (for 1932) of 202 mph. The machine spanned 32 ft and was 26 ft long. The later 17L introduced a retracting undercarriage and there were many sub-variants. More than 780 Beech 17s were built and some were still flying in the early 1970s. But Beechcraft's big fortunes were made by their D18 light twin monoplane, an eight seater originally intended for feeder liner work and announced in 1937. This Beechcraft model had a 47 ft span, was 34 ft long and two 350 hp Wright radials gave the 18A a cruising speed of 205 mph. The Beech 18 became popular not only with small airlines, but also with companies seeking executive transport. Over 5,000 were built as military trainers and transports during the Second World War, and it has remained in service ever since.

The roll call of companies active in the American light plane market of the 1930s included many illustrious names, some of which failed to survive the Depression years, while others were producing aircraft until after the Second World War and were gradually absorbed by the 'Big Three' – Piper, Beech, Cessna – or simply went to the wall in the face of the 'Big Three's' domination of the market. Among the big names were Aeronca, whose C-3 of 1931 was an early light plane success which was produced until 1936, some being built in England. Succeeding Aeronca models were produced until 1950 when the Sedan was the last type built. Bellanca was a firm who got into aviation by building a special aircraft to demonstrate the first 200 hp Wright radial engine in 1924. Later they built some successful light and utility aircraft of which the Pacemaker, Skyrocket, and Air-cruiser of the early 1930s were the most successful. Other notable names were Fairchild, Fleet, Mooney, Luscombe – whose Silvaire series sold around 8,000 before and after the Second World War – Porterfield, Rearwin, Stearman (absorbed by Boeing in 1930), and Waco.

One firm whose output reflected the sheer confidence of the American light aircraft industry in the 1930s was Howard. Ben Howard chose the designation DGA for his models, eg Howard DGA-15, the last of the line. On enquiring the significance of DGA in the title, the prospective purchaser was told 'Damned Good Airplane'! Howard did well in the 1920s with an early model, the DGA-2, a fast little biplane said to have been popular for running in bootleg whisky in Prohibition days.

In the 1930s light aviation became a commonplace form of transport in USA. Virtually every township of any size had its landing strip, and the industry thrived, largely for the vast home market but doing well in export markets too. There was plenty of air space in North America and the distances involved between commercial centres made light aircraft a genuinely economical form of transport for businessmen as well. This led to the growth of air taxi firms and the use of executive light aircraft, owned and operated by companies. There was no government involvement and the market grew freely. Sports flying was popular, with 'fly ins' and local air displays as common summer attractions.

Air racing became a pastime with a big spectator following, and it bred its own type of aircraft, built for speed and performance and quite distinct from the conventional type of light aircraft. This was in contrast to Britain or Europe where modified, or even stock, production machines were used for racing, except in such 'international' events as the Schneider Trophy series. The American racers, however, undoubtedly improved the breed, and the excellence of the fighters available from America in the 1940–42 period was in no small measure due to experience in building better and sleeker racing planes in the twenties and thirties. Strictly speaking it was US Navy and US Army funds that got air racing going in America in the form it most commonly took.

A Pulitzer Trophy Race series from 1920 onwards led to a

series of fine Curtiss Racers (R-6 for the Army, R2C-1 for the Navy) with speeds of over 230 mph. The famous Curtiss Hawk series of biplane fighters were developed from these sleek racing machines. Meanwhile, the Gordon Bennett Trophy Race of 1920 saw the first aircraft with fully retracting undercarriage, the Dayton-Wright RB with fully enclosed cockpit and high wings. Other major annual races were the Thompson and Bendix Trophy series. The tubby and distinctive Gee-Bee Sportster and Super Sportster were made famous by Jimmy Doolittle in the Thompson races of the early 1930s. The Super Sportster Model Z had a 535 hp Wasp engine and won the 1931 Thompson race at 236 mph. In 1932, the Super Sportster Model R-1 had a 800 hp Pratt and Whitney Hornet engine and won the Thompson race at 252 mph. The French took an interest, too, and in 1936 entered the sleek Caudron C-46 which won at 264 mph, and also held the speed record of 314 mph. It had a 340 hp Renault, air-cooled, in-line engine. Howard Hughes was a contemporary who built a fighter-like Hughes Racer in 1935 with a 700 hp Twin Wasp radial engine. This machine later held the world speed record with 352·46 mph. In 1937 he flew across America in $7\frac{1}{2}$ hours in the same aircraft.

All these activities were a far cry from light aviation, however. Easier to grasp for the man in the street was one headline-hitting achievement, that of the Lockheed Vega cabin aircraft 'Winnie Mae' which Wiley Post and Harold Gatty flew round the world in only eight days in 1931, via Berlin and Moscow. The Lockheed Vega was, in fact, the first Lockheed aircraft and was a neat, single-engine, high-wing type.

Light aircraft development in Germany was rather different after the First World War. Following the 1918 Armistice, the Treaty of Versailles imposed harsh terms which included the prohibition of manufacture or import of powered aircraft.

Arrow Active

above, right
A gliding club in Germany in
the 1930s

above, left
Members of the German Air
Sports Organization prepare a
sailplane for flight in 1938

Gliding was not affected by the ban, however, and this was
to make Germany the leading nation in design and development
of sports gliders. Other nations in the 1920s were too
pre-occupied with developing powered machines to bother much
with gliding. Gross Wasserkuppe, excellent downland country
north-east of Frankfurt, became the centre of activity. In August
1921 one of the keener gliding enthusiasts, Wolfgang Klemperer,
a former air force pilot, made his name by staying airborne
for over 13 minutes, then a world record, in a glider of his
own design, the 'Blue Mouse'. In a matter of days, other pilots
broke this record, but it served to show the potential in glider
development and raised enthusiasm to a new pitch. A year later
one-hour flights were achieved.

Some standard glider types were built; there was a Zögling
(primary) glider, the Kegel Prüfling (secondary) glider and some
fine high performance sailplanes such as the Professor, Schulz,
and Espenlaub. One of the most remarkable types was the
Schwabe, which was little more than a refined, long-wingspan
Zögling. In 1929 it took the gliding duration record to 14 hours
43 minutes. In the late 1930s one of the finest duration sailplanes
was the beautiful, gull-wing Minimoa. Other gliding centres in
Germany grew up at Rossitten, on the Baltic, and Munich.

Soon Germany became the Mecca for gliding enthusiasts and
in the mid-1920s the craze had spread to France, Poland, and
Russia, Germany's nearest neighbours. By 1930 it was
established in Britain and America as well. The famous London
Gliding Club was formed at the end of 1929 and by the following
summer the club was establishing its fine flying site at Dunstable
Down, with a handful of German gliders for their first attempts
at gliding. In Britain Kirby and Slingsby built a successful series
of gliders, both before and after the Second World War, starting
in the early 1930s. Early American gliders included the
Baker-McMillan Cadet, the Boulus Albatross and Braley
Skyport. The last two were modelled closely on German sailplane
and Zögling types respectively. More unusual was the Heath
Super Soarer, a biplane glider which in 1930 became the first
ever unpowered machine to loop the loop, flown by its designer,
Ed Heath.

When Hitler's Nazi party came to power in 1933, the already
well-established gliding movement in Germany was seen as an

top
The sport of gliding—Peter Scott's Slingsby Eagle is towed up by a Tiger Moth during the Gliding Championships at Lasham in 1957

right
An Olympia 460 sailplane of 1960

ideal way to train youngsters for flying in the planned build-up of the German airforce. Gliding in Zögling machines became part of the syllabus for many Hitler Youth movement summer camps, while a complete 'shadow' flying and gliding organisation, Reichsluftsportverband (German air sport organisation) was established to administer sports flying and training.

By 1935, seeing developments in Germany, the British government was persuaded to give modest subsidies to leading gliding clubs. Gliding subsequently became part of the Air Cadet and Air Scout training schedules, as in Germany, and in most major countries since the 1930s a similar philosophy of introducing youth to airmanship has prevailed. Gliding now, as in the 1930s, has remained a major international sport and the development of gliding technology continues, though the pace of development is inherently slower than the technology of powered flight. Gliding records for height, duration, and distance have been continually broken and have reached astonishing figures. By the 1960s, for instance, the single-seater distance record was nearly 650 miles and the height record, 46,000 ft.

Helicopters

An alternative notion to flying with wings had occupied the attention of many of the pioneer theorists of flight. Lift in the kite style was one thing, but what was clearly observed centuries ago was the 'windmill effect'. Large, paddle-shaped sails held into the wind and pivoted would turn and generate power. In Europe this power was most often utilised to turn millstones or raise water. Post-type windmills of the traditional type–where the sails could be turned to face the prevailing wind–are known to have existed in the thirteenth century, the period of the earliest existing illustration of such a mill. But it seems probable that such windmills were known and used for centuries before that. In the Middle East a popular type of windmill had its sails set horizontally like the rotors on a modern helicopter. The child's windmill was for many centuries a favourite toy, long before any aircraft ever flew, and a variation of this toy, dating at least from the fourteenth century, was the string-operated windmill whereby a string wrapped round the spindle was tugged away to cause the rotor to spin. It demonstrated hundreds of years before man flew that imparting power to a rotor would make it spin and cause lift–and the lift effect (a tugging action) will be evident to anyone who has ever held a toy windmill in a strong breeze.

Leonardo da Vinci included a type of horizontal rotor flying (or rather lifting) machine in his papers, though it was little more than a concept showing how a spinning helix could theoretically give a vertical lift-off. In April 1784 a model was demonstrated by two French inventors, Launoy and Bienvenu, which was a primitive form of helicopter. Two two-bladed paddles (or propellers) were arranged at the opposite ends of a spindle so as to contra-rotate. Power was provided by a bow-string which was wound up and released, spinning the two paddles to give vertical lift.

In 1828 an Italian, Vittorio Sarti, produced a model of a much improved version of the same idea. Here there were two four-bladed rotors with properly angled, windmill-like sails. The rotors contra-rotated with one spindle arranged inside the other. A steam generator in a gondola at the bottom was supposed to provide the power by directing jets of steam at the rotors, and a large triangular sail suspended from the gondola was provided as a stabiliser or rudder. The vague part of this proposal was the mode of drive. Like all the early ideas for aircraft, lack of an efficient and light power unit was a major obstacle to practical development. Sir George Cayley, already noted for his major contribution to winged flight, was the man who published the first really practical idea for a helicopter in which

all the essential features of the modern machine were to be seen.

Cayley had earlier made his own version of the Launoy and Bienvenu model helicopter. His actual design is said to have been inspired by a simpler idea sent to him by an American enthusiast called Robert Taylor. Once again, however, lack of an efficient power unit prevented this fine machine from being built. Drawings and details of Cayley's Patent Aerial Carriage were published in *Mechanics Magazine* in April 1843. A gondola with tricycle wheel undercarriage held the steam engine and aviator. It was decorated with a bird's head at the front and a bird-type horizontal tail at the rear to act as a stabiliser. Two contra-rotating rotors paired on common spindles were outriggered on each side of the gondola, with a belt drive from the engine. Two pusher propellers for horizontal flight were driven by belting from the engine, and a vertical, fabric-covered rudder was erected between the propellers. Cayley's machine was never built, but all subsequent helicopters followed the basic principle which Cayley postulated.

A year before Cayley's design was published, another Englishman, W. H. Phillips, built a working model helicopter with steam jets from the rotor tips. The steam was generated by charcoal and gypsum. Vicomte Gustave d'Amecourt made some helicopter models in France in the 1860's, and he invented the name 'helicopter' from the Greek *helicos* (a spiral) and *pteron* (wing).

In 1871 Pomes de la Pauze in France produced a design with a large twin-blade rotor and a vertical stabiliser. There was provision for altering the pitch of the rotor in flight as on modern helicopters. There was no suitable motor, however, and the machine was not actually built.

Charles and Louis Breguet in France built the first helicopter which actually flew in September 1907. The Breguet machine had four rotors and was flown by their assistant, Richet. The flight was nominal, though, with four men holding the aircraft steady while it lifted itself a short way into the air. Another French aviator, Cornu, was actually the first to fly freely in a helicopter only two months after Richet, in November 1907. His was a two-rotor machine with belt drive and the 24 hp motor mounted in a pram-like body. His aircraft rose to a height of 6 ft! In 1908 the Breguet brothers built a new machine which they called their Gyroplane Nr.2. This had two $25\frac{3}{4}$ ft diameter tilting rotors and wings. In July it made a promising flight to a height of 15 ft and then moved 60 ft forward. In a later flight it was damaged in a crash, and was later destroyed in a storm, thus ending the Breguet experiments for the time being.

A big portent for the future came when a young engineer from Kiev, Igor Sikorsky, built a small helicopter with twin rotors in 1909. Though it ran successfully it could not generate enough lift to get airborne. In 1910 Sikorsky tried again with a second similar but bigger machine. The engine drove the rotors via belts and shaft. Though this machine made one or two hovering flights while tethered it was scarcely able to lift itself off the ground. Sikorsky lost interest at this point and carried on working on conventional aircraft design. After the Bolshevik uprising of 1917–22 Sikorsky left Russia and settled in America where he was to become the greatest influence of all on modern helicopter design.

A Danish engineer, Jacob Ellehamner, built a model helicopter in 1911 which was so successful that he made a full-size version with a single, large, contra-rotating rotor and a tractor airscrew. Between 1912 and 1916 he made several flights, but it finally crashed and was written off.

A Spaniard, the Marquis Paul Pescara, was responsible for the first fully working helicopter and he realised that lightness was all. He had the bright idea of using a Gnome rotary engine to drive the rotors. This worked well but there was little directional control or stability. The following year Pescara made another helicopter in France without further improvement. In 1923 came the breakthrough. He made yet another machine, this time with a lever arrangement enabling the pitch of the rotor blades to be changed in flight. The rotors on these Pescara helicopters were of the braced, fabric-covered, biplane type–each rotor like a miniature biplane. The rotor could also be tilted in flight, and a clutch was provided to disengage the rotors should the engine fail. Thus disconnected, the rotors would 'autorotate' in the airstream thus lowering the machine reasonably gently to earth rather than allowing it to plummet to the ground. Autorotation of the rotors has been a standard helicopter feature ever since. Pescara now had vertical flight capability and a good degree of control in horizontal flight–indeed, the variable pitch and tilt facility is also now a standard helicopter design feature. On 1 June 1923, Pescara made his first properly observed flight, climbing to a height of 6 ft, then flying horizontally for 273 ft. He soon bettered this initial performance. By January 1924 he had flown 1,640 ft, turned and flown back.

By now Pescara had a French rival, Étienne Oemichen, a motor car engineer from the Peugeot concern. He built several rotary-engined helicopters and by late 1922 had achieved a flyable design. This had two rotors and eight propellers. Two propellers gave forward flight, one steered, and five maintained stability. The two rotors were twin bladed. Steel tubing was used for the airframe construction. Oemichen carried on improving his design and in May 1924 he made the first helicopter closed-circuit flight, 5,550 ft at a height of 50 ft. Later that year he set up the first helicopter lift records with cargo payloads of 100 and 200 kg lifted to a height of 3 ft! At this time Pescara held the world helicopter speed record of 8 mph! Oemichen, however, had by now eclipsed the efforts of Pescara, who withdrew from the scene.

In America Dr. George de Bothezab was the most successful pioneer. He built an X-shaped machine ('Flying X') and fitted four big rotors, one at each corner of the X. This machine made

Cierva's C–5, of 1925, was an improved version of his 1923 C–4 and of similar appearance

vertical flights of 6 ft. It was tested by the US Army who rejected it as too complicated. There were other less successful helicopter pioneers. In England Louis Brennan built a machine in 1924 with a 60 ft diameter rotor turned by propellers in two of the blades which were driven by shafts inside the blades. It made 75 short flights, then crashed.

Meantime vertical flight was being developed in a subtly different way and the instigator here was a Spaniard, Juan de la Cierva. He had been a glider builder before the First World War and the tendency of aircraft to stall—lose lift due to lack of airspeed—and then drop, caused him to look at ways of obviating this. Cierva's idea was to take a conventional fuselage with front engine and airscrew, then to place a pylon on the fuselage which supported a large diameter, free-wheeling rotor. Forward airspeed was given by the conventional power plant. As the machine gathered ground speed the rotors started to windmill in the airstream and so provided lift to keep it airborne. Cierva's first flights, with his C-4, took place in January 1923. Most important of all, Cierva overcame the problem which had dogged the earlier rotor craft. With fixed rotors the pitch of the blades is constant. As the rotor rotates the forward rotor blades move faster through the air than the retreating blades and the forward blades thus tend to lift faster than the retreating blades. This caused the machine to become unstable and rock. Cierva perceived that by allowing the rotor blades to flap loosely on the rotor they would change their relative angles of pitch as they rotated, thus balancing out the tendency to lift more while forward than retreating. This fundamental idea very largely solved the stability question, and years later the same facility was given to helicopter rotor blades.

Cierva now went from strength to strength. His C-6A was built on the fuselage of an Avro 504N and in 1925 Cierva took it to England. The Air Ministry liked it and asked the British Avro firm to build some machines of the Cierva type. Cierva called his type of aircraft an 'autogiro', a name since accepted worldwide. The autogiro would not normally take off without a short forward run to get up rotor speed and lift. However, Cierva made quite successful attempts to get direct vertical take-off. In his C-30 design of 1933, he clutched the rotor shaft to the engine. The rotor blades were set at zero pitch and run up at speed. The pilot then declutched the rotor from the engine, threw the fast moving blades into pitch, and the aircraft jumped instantly into the air as the properly pitched rotor blades gave immediate lift. Once airborne, the tractor engine drew the machine forward in the normal way.

Autogiros became quite common in the 1930s. The RAF had a squadron of them, one fitted with floats. The Avro-built machine was called the Rota, and civilian registered aircraft were used by the police and the post office as well as by private flyers. In 1933 a Cierva C-40 was flown on and off a Spanish freighter at sea, and in a more spectacular trial one landed and took off from an Italian cruiser at sea. In the USA Pitcairn built Cierva models under licence, as did Kellet. The US Army purchased one or two for evaluation and ordered more. In 1939 the US Mail used a Kellett KD-1 to run the first air mail autogiro service between Philadelphia and Camden, N.J. As it happened, however, this promising start to the career of the autogiro was not sustained. When the Second World War came along, it was the helicopter which took over the vertical take-off story. Though one or two autogiro types have been designed since 1939, the commercial autogiro was eclipsed by the helicopter, for this type proved to have far more development potential, and more important still, control was better and it could lift and carry superior payloads.

In the field of helicopter development the Frenchman Louis Breguet returned to the scene in 1929, some 20 years after his first attempts with his Gyroplanes had failed. He formed a company to build and exploit a new design which he called the Gyroplane Laboratoire. This had an open, steel-tube framed body with a plywood-covered, aircraft-type tail; there was also a tricycle-type undercarriage on wide outriggers. The 350 hp Hispano radial engine drove two 52 ft diameter two-bladed rotors which contra-rotated on two concentric shafts. The rotor blades were hinged. Full pitch control was featured, and the pitch of the two rotors could be varied relative to each other so that the aircraft could be turned to the left or right. The new machine was built in 1933 and made its first free flight in June 1935. It proved highly successful and broke nearly all the then existing helicopter records by the end of 1936 – closed circuit of 1,640 ft, height of 518 ft, forward speed of 67 mph and endurance of 62 minutes in the air were among the successful records

The Avro Rota autogiro of the Royal Air Force, 1936

claimed, and the designer Breguet thus had his visionary ideas of 1909 justly vindicated by his own aircraft nearly 30 years later. Work on improving the Gyroplane Laboratoire continued until the Second World War broke out in September 1939. It was stored in a hangar at Villacoublay, completely neglected after the fall of France, and was destroyed in a bombing raid. Often overlooked by air historians, the Gyroplane Laboratoire can lay claim against German, Soviet and American rivals of the time to be the first practical helicopter to fly successfully.

Across the border in Germany, Breguet's efforts were being rivalled by Dr Heinrich Focke, founder of the Focke-Wulf aircraft company. Focke-Wulf had the German licence to build Cierva autogiros (just as Avro had the British licence) and Focke used the Cierva layout as his starting point. He wanted to drive the rotors, however, to improve on the limitations of the autogiro system. He was working on the problem in 1933 when the Nazi party came to power and had Focke thrown out of the company for his unsympathetic political views. Undeterred, Focke set up his own company, Focke-Achgelis, to carry on with his design work. His helicopter, the Fw 61, was built as a flying model first in 1934, flew successfully, and was then built full-size. Like Cierva and Avro, Focke used a training plane fuselage and (the Fw 44) tail and even retained the original, front-mounted radial engine. The tractor propeller was retained, cut down to stubs at the spinner, and shaft drives from the motor drove the two rotors which were mounted on tubular steel outriggers which also carried the side wheels of the tricycle undercarriage.

The Fw 61 prototype made its first flight a full year after Breguet's machine in France. In 1937, however, it made a big impact when it smashed decisively all the helicopter records held by Breguet. By mid 1938 it held the distance record of 141 miles and had reached a record height of over 11,200 ft. A noted woman pilot of the day, Hanna Reitsch, got good publicity worldwide by flying the Fw 61 indoors, in a Berlin covered stadium. After two Fw 61s had been built Focke was asked to build a large six-passenger version, the Fa 266. The British Air Ministry was impressed and tried – unsuccessfully – to buy a Fw 61 for evaluation, but the outbreak of war changed the situation and Focke was diverted to more important war work.

Soviet Russia was the home of much work on helicopter design in the 1930s, the government having set up a special design section in 1928. The first Russian helicopter was the I-EA of 1930, which was quite successful. It was an open framework

bottom
The Sikorsky R-4 in RAF service (as the Hoverfly) in 1945, the first operational military helicopter

below
Igor Sikorsky at the controls of his VS-300 prototype on its first tethered flight in September 1939. Later it was given a fabric-covered nose

machine with a single, large rotor and propellers fore and aft. By 1936 improved versions were flying, but the infamous Stalin purges decimated the design team and slowed up development. However, the last model, 11-EA, was completed and test flown in 1940–41.

A successful helicopter was also built in Britain, designed by Cierva's chief designer. Built by G. and J. Weir Ltd, the machine resembled the layout of the Fw 61 but had an open, steel-frame tubed fuselage. It was designated the W.5 and first flew in June 1938, making many more flights to overcome initial 'teething troubles'. A bigger machine, the W.6, was produced which could carry two passengers. It first flew in October 1939 and crashed soon afterwards. Repaired, it carried out a successful test programme, but war priorities intervened and work on these promising designs ceased.

In the United States, Sikorsky, like Breguet, returned to his old love, helicopters, after 30 years. His interest had been revived in 1928 and he worked out some notional designs. In 1938, having noted progress in Europe, he persuaded his employer, United Aircraft, to build a prototype for a new helicopter design, the VS-300. It flew, tethered to the ground and piloted by Sikorsky himself, in September 1939. The design was essentially that which has characterised helicopters ever since, and the success of the VS-300 emphasises that though Sikorsky may not have been *the* pioneer helicopter designer, he was the most important one. The VS-300 had a metal tube frame with an insect-like fuselage, open cockpit, and fabric covered at the front end. A 75 hp, air-cooled engine drove the main rotor. First free flight was made in May 1940, but there were stability and control problems. After trying three tail rotors at various angles, Sikorsky found that one single tail rotor on a metal tube outrigger solved the problem. Except in a few designs, that is where the small tail rotor has remained ever since. By May 1941 the VS-300 was flying in its modified form, and it then started to beat all the records held by the Fw 61. By now war seemed likely, and the US Army asked Sikorsky to design a helicopter specifically for military use. This emanated in 1942 as the Sikorsky R-4, a long, dragonfly-like aircraft which resembled its progenitor, the VS-300 in shape, but had the added refinement of an enclosed pilot's cabin, two seats and dual control. Top speed was 82 mph, maximum operating height 8,000 ft and range 220 miles.

The R-4 was put into immediate production and by 1944 was coming into service in Europe and the Pacific areas. Aside from the US forces, R-4s went to the RAF and the US Coast Guard. It was with the latter service that an early demonstration of the helicopter's potential was given. A destroyer suffered a serious explosion off the New Jersey coast and the R-4s of the newly-formed Coast Guard detachment were used successfully to land injured men and carry out medical supplies to the ship.

During the Second World War, apart from the American R-4, there was limited but interesting helicopter development. Japan acquired a Kellett autogiro in 1939 from the American licence builders of the Cierva helicopter, and a close copy was made, the Ka-1. Over 240 were built.

Major development took place in Germany, under the impetus of war. Friedrich von Doblhoff was a light aircraft designer with an Austrian firm and he theorised that helicopter performance could be greatly improved by motors at the rotor tips. Combustion jets seemed the easiest answer, and in 1942 a test rig was built to test the theory. A three-blade rotor had small jets which were fed through the rotor blades with a mix of petrol vapour and compressed air, and ignited with a spark plug. The idea worked and Doblhoff was given a research grant to perfect a prototype. The first result was a small observation helicopter, with WNF-342, which was intended for use on submarines. U-boats suffered the tactical disadvantage when surfaced of having only a limited view and the helicopter was an idea to enable the U-boat commander to see beyond his immediate horizon when hunting for convoys. In August 1943 the prototype was destroyed when Doblhoff's Vienna workshop was bombed. A second machine was quickly built, but the rotor jets proved to be very heavy on fuel. In subsequent prototypes, therefore, the jets were used only for take-off and landing, with a radial engine and rear pusher propeller for horizontal flight. The compressed air system for the jets was driven from the engine. More prototypes followed, but when the war ended in 1945 no production had started. The surviving prototypes were taken to USA and some of the ideas were used in post-war designs with rotor-tip jets.

In the meantime, another firm, Flettner, headed by Anton Flettner, produced a helicopter, the Fl-265, for U-boat service. The Fl-265 was of conventional drive with a 150 hp radial engine and contra-rotating rotors. This type, with a 99 mph top speed, had successful trials from both surface vessels (as a submarine searcher) and from U-boats in 1940. It was also tested as an Army 'spotter' plane and as an early type of aerial crane, lifting assault boats. An improved design went into production, the Fl-282 Kolibri, with a 150 hp engine, though relatively few saw action since the factories were bombed in 1944. Those Fl-282s which were delivered, however, saw extensive service at sea on surface vessels—though the use of helicopters on warships is often thought of as a post-war idea, it was actually pioneered by the Kriegsmarine in the 1940–43 period. Yet more Flettner designs were on the drawing board by the time the war ended, including a bomb-carrying version of the Fl-282 for fleet service, and the Fl-339, a utility personnel and cargo carrier. Also produced was a tiny, folding, 'kite' helicopter with a motorcycle engine, specifically as a U-boat aircraft.. It flew tethered, with the airstream created by the speed of the U-boat assisting in maintaining lift. This machine was small enough to stow in the casing of the submarine. It was intended to fulfil the function of the original Kriegsmarine helicopter requirement, to extend the horizon of the submarine commander.

A feature of all the Flettner designs was inertia damping of the controls and some key components to overcome the problems caused by heavy vibration which was always a characteristic of helicopters, especially the early ones. After 1945 Flettner went to America where he continued to design and build helicopters.

Dr Focke and his Focke-Achgelis firm also built military helicopters. In 1938 the German Navy requested a helicopter big enough to patrol convoys, carry torpedoes, and lay mines—about 15 years before any other navy got around to the same sort of idea. Focke adapted his passenger-carrying helicopter, the Fa 266, to meet this requirement, under the designation Fa 223 Drache (*Drache:* Dragon). In layout, the Fa

223 was similar to Focke's earlier designs, with rotors on outriggers. It had an enclosed cockpit for two side-by-side pilots, plus a rear compartment seating four, and the 980 hp BMW engine right aft. Top speed was 114 mph (124 mph in later models) and the range was about 200 miles. Tests in 1942 were successful, but bombing of the factories disrupted production severely and very few were completed. There was also a version of this aircraft, the Fa 223E, for military use, with a supercharged engine boosting horsepower and speed. In exercises the Fa 223E was used to lift and carry light vehicles and stores, again some years before this form of military, battlefield, helicopter support is generally considered to have been evolved. Clearly, had German industry not been so badly affected by Allied bombing, the helicopter would have come to play a larger part in land and sea warfare on the German side in the Second World War than it ever actually did–the foresight, plans and will were there, but the aircraft were in short supply.

Focke designed a small, non-powered, 'kite' glider for submarines, simpler even than Flettners, and this was given tests but was not produced. The lift was obtained solely by towing the aircraft, causing the rotors to revolve. Other Focke projects included a huge flying crane able to lift interchangeable cargo carriers. This was the Fa 284, and though it was never built, some aircraft of similar design were built and flown in other countries (such as America and Russia) 10 or more years later. They were then hailed as innovations, but they were actually vindicating the much earlier ideas of Dr Focke. Also designed by Dr Focke was the Fa 225 helicopter glider–it had a standard DFS 230 glider fuselage, with a helicopter rotor. The idea was that this glider could be released to land on very short strips (60 feet or less).

It was after the Second World War, however, that the helicopter came fully into its own. After Sikorsky's and Focke's original ideas had been studied and absorbed, helicopter technology began to advance quickly, and new designs streamed from technical departments. Helicopters cannot be said to have revolutionised either air transport or warfare, but they made a powerful contribution to its development after 1945.

Much of the development was to meet military requirements. The helicopter became such an essential part of the modern military scene that it becomes difficult to remember how the navy and army commanders of the Second World War ever managed without them. With naval forces, the helicopter of post-war years was used as a personnel transport, and more particularly as a 'plane guard'–a rescue aircraft for aircraft-carrier operations where the accident rate was relatively high and pilots of crashed aircraft had to be rescued from the sea. In pre-helicopter days an accompanying destroyer had to perform this rescue role, but it was quickly usurped by the helicopter. Nets, scoops and winchropes were the most common ways of picking up men from the water. In latter years rescue helicopters have often carried divers who jump from the helicopter into the sea to assist the men being rescued. By the time of the Korean War in 1950, the helicopter was already established in this naval role. It soon showed how it could be used to rescue downed pilots from behind the enemy lines, provided it could be vectored to the scene. There were legendary stories of fighters pinning down enemy troops long enough for

The Focke-Achgelis Fa 330 was the non-powered 'kite' helicopter glider for use from submarines

a helicopter to arrive and pick up the crew of a crashed aircraft.

The idea of torpedo carrying, minelaying and, indeed, mine-sweeping by helicopter came in the late fifties and sixties. Modern navies had frigates and other major ships in service with helicopters for the anti-submarine role, the helicopters carrying homing torpedoes for attacking submarines. The other big innovation, revolutionising anti-submarine warfare, was 'dunking sonar'–a helicopter carrying a sonar (or asdic) transponder which could be lowered into the water while the operator in the helicopter listened for submarines just as he would in a ship. Several helicopters could pattern themselves over a wide sea area to search for submarines. Warships with torpedo-carrying helicopters could then be summoned in for the kill.

The concept of 'vertical envelopment' also involved the use of the helicopter. First tried on a big scale by the British in 1956 at Suez, this involved transporting troops by helicopter force from their troopships or special carriers direct to the beachhead or attack point–with returning helicopters evacuating the wounded. Heavy weapons can also be carried, either inside or slung beneath. Landing troops from ships by helicopters is now a major aspect of amphibious warfare and some powers have complete Marine units organised for this kind of operation. A special type of ship–a 'mini' aircraft carrier, with helicopter-only facilities–came into service in the 1960s for amphibious warfare operations of this type.

Allied to helicopter operations from the sea was the 'air cavalry' concept. Complete helicopter-borne units could be rushed into battle as required by the army commander–or lifted from trouble spot to trouble spot. The ultimate development of this idea was the US 1st Airborne Cavalry Division, deployed in the Vietnam war. This type of division has complete air support, with troop carriers, recce and armed support helicopters. The armed support helicopter or 'gunship' was very much a product of the 1960s and the United States involvement in the Vietnam war. This was a limited war operation where helicopters could operate with reasonable impunity since the United States had command of the air, almost a prerequisite for unlimited helicopter operations as the slow moving 'chopper' is particularly vulnerable to small missiles and gunfire.

In commercial, passenger-carrying service, the helicopter has never seen such extensive use as in its military role. This is mainly due to the high operating costs. Fuel consumption is high, helicopter endurance is, in general, shorter, and the carrying capacity is limited in most. There is also the noise and vibration problem which only began to be solved by the most advanced of the 'second generation' designs. Thus few scheduled passenger services have been able to show as much profit as services operated by feeder liner type aircraft–if, indeed, they have shown a profit at all. So the helicopter in commercial use has tended to be employed for very specialised services–search and rescue by civil authorities, air taxi or charter, surveying, servicing oil rigs or lighthouses or short distance cargo work. Only in relatively recent years have helicopter designs advanced sufficiently to make short distance feeder liner work feasible. Notable services are those operated by New York Airways between the various New York airfields and the roof of the Pan Am terminal in Wall Street, New York, and the British Airways service between Penzance, Cornwall and the Scilly Isles.

Helicopter development since 1945 has been rapid and most of the running has been made by the USA. The Sikorsky firm followed their first production type, the R-4 (known in Britain as the Hoverfly) with the R-6 which was a streamlined, all-metal derivative of the fabric-covered R-4, and was produced in 1944. The first helicopter to become really familiar to the layman, however, was the Sikorsky S-51 which appeared in 1946 and was in production later that year. Known to the British as the Dragonfly, over 300 were built in Britain under licence by Westland Aircraft Ltd. The S-51 was a four seater with a handsome 'glasshouse' cabin and clean lines. Serving the US Air Force as the H-5 and the US Navy and Coast Guard as the HO-3S, the S-51 was the helicopter which made its name in the Korean war of 1950–53 and thereby made a most valuable contribution to military helicopter development. The S-51 was the machine which, operating from American and British aircraft carriers, rescued many a crashed pilot from the icy sea or from far behind the enemy lines – before the days of helicopters, such an involuntary landing in enemy territory would have meant certain capture. Techniques were developed whereby aircraft in company with a crashed plane swooped low to hold off any soldiers attempting to capture the downed pilot while a helicopter was summoned to pick him up. The S-51 was also used for casualty evacuation from the front line to mobile field hospitals, and soon almost supplanted the Army Stinson and Piper light airplanes which had been intended for the casevac role. S-51s were also the first helicopters used by civil concerns. In 1947 they were used to start an airmail service in Los Angeles, the first scheduled helicopter service in USA and in 1950 British European Airways started the first scheduled helicopter passenger service in the world, between Cardiff and Liverpool.

In 1949 came an enlarged helicopter from Sikorsky, the S-55, which had the engine moved to the nose, so making available maximum fuselage capacity for the payload. Numerous models were developed, including a gas-turbine powered version by the British licensees, Westland. In Britain the S-55 was called the

below, left
The Sikorsky S–55 was a classic 10-passenger helicopter built in many forms from 1949 on. This is the US Army H–19 version, in service in the Korean War in 1951

below, right
The Sikorsky S–56, in service as the Mojave, was the first specialized assault helicopter able to carry military vehicles and guns

Whirlwind, to the USAF the H-19, and to the US Navy the HO-4S. It carried up to 10 passengers, but was also produced as an anti-submarine machine with dunking sonar, an anti-submarine attack aircraft (in the US Navy) with torpedoes, a troop carrier, a rescue aircraft with winch, and for several other specialised roles. Some models were fitted with floats. In civil guise it was a popular machine for charter work but was also used by several airlines, including BEA in Britain, New York Airways, and Sabena in Belgium, the latter service (in 1953) linking several European capitals. The original S-55 was powered by a 600 hp Pratt and Whitney Wasp engine, but there were numerous variants. Several hundred S-55s of all types were built and many were still in service in the 1970s.

A big jump forward came with the S-56 of 1953, which was the biggest helicopter of its day. It was a specialised assault machine for the US Marines, the first of its kind, and carried 26 troops or three jeeps. A Double Wasp engine of 1,900 hp was carried in a nacelle on each side. This left the fuselage clear for cargo, and there were 'clamshell' nose doors for loading. It had a 90 ft diameter five-blade rotor–it could fly with one blade shot away–and was 59 ft long. Known as the Mojave, it served both the US Army and US Marines. The Sikorsky S-58 was another classic type, essentially a lengthened and more powerful version of the S-55. Initially it had a 1,525 hp Wright Cyclone radial engine, but the British Westland-built version, the Wessex, improved on the original design by having a Napier Gazelle gas turbine engine. Well over 2,000 S-58s were built.

The S-61 of 1959 was the next outstanding Sikorsky design, and has generally been acknowledged as one of the finest helicopters of all time. It has been developed for numerous special purposes and in its various forms it is widely used for both military and commercial roles the world over. The British version, licence built by Westland, is called the Sea King. To the US Navy it is the SH-3, and as the S-61N it serves numerous airlines and charter operators. The SH-3 became particularly well known as the helicopter which took a prominent and important part in the recovery of returning American astronauts from the sea. The anti-submarine and rescue versions were particularly sophisticated designs with computer-assisted navigation equipment and provision for 'flying' the aircraft from the aircraft side door by means of small over-ride controls to allow the aircraft's hovering position to be adjusted precisely for rescue purposes. The S-61 in its military form also had an all-weather capability. A purely cargo version with loading ramp was also produced.

A French Djinn light helicopter on rescue service in the Alps

More dramatic than the S-61 was its contemporary, the S-60 Skycrane. This was developed into the S-64 and, as the CH-54, it entered US Army service in 1964. This was literally a flying crane, the concept being that the basic aircraft could lift a standard wheeled pod or container unit, deposit it at the desired destination, fly back for another, or return with an 'empty'. The standard pod could carry cargo or passengers, and further adaptations saw pods in use as command centres, first aid posts, or ambulance units. The maximum number of people lifted in one pod was 87, though it would generally be 68. When not used with pods the CH-54 could lift light tanks, dozers or wrecked aircraft up to 20,000 lb in weight. The aircraft became standard equipment with the air cavalry divisions.

above
The Sikorsky S–61 N commercial helicopter operating a passenger service to the Scilly Isles

top
The Sikorsky S–58 was widely used. This is the British-built Westland Wessex version of the Royal Navy carrying troops into the Borneo jungle in 1966 on anti-guerrilla patrol

With these designs and others the Sikorsky company had a fair claim to be the world's leading helicopter design and production firm, but other American concerns also have fine records in the helicopter field. Outstanding is the Bell company, whose little Bell 47 utility helicopter is one of the most widely used in the world. In its various forms it has the longest production record, having been built continuously since 1946. Well over 4,000 were built in the first 20 years of production and almost every country in the Western world has examples flying. It is a standard service light aircraft in USA, Britain (as the Sioux), Australia, New Zealand, Brazil, Spain, Turkey, Italy, Western Germany, Argentina, and many other countries. It is also a popular police helicopter and numerous examples serve large industrial concerns (like oil companies) and air charter firms.

The first Bell helicopter, the Bell 30, flew in 1943, and the Bell 47 appeared in 1945. Its structure is distinctive–an uncovered, metal-tube 'skeleton' fuselage with Plexiglass 'bubble' cockpit and a 200 hp Lycoming or Franklin engine, fully exposed, in most models. There is a skid and wheel-type undercarriage. Later models, mainly intended for the civil executive market, were more refined, an example being the Bell 47H which has full metal cladding over the entire fuselage giving a much more streamlined appearance.

Two other, particularly successful Bell designs were the UH-1 Iroquois and the AH-1 Huey Cobra. The Iroquois was in production in the early 1960s as a general utility helicopter for service use. It was essentially a stretched Bell 47 as far as general layout goes, with room for 7–10 troops, or three stretchers in its cargo compartment behind the pilot. Top speed was 138 mph. In Vietnam the need for gunfire support as an integral part of a helicopter assault force led to the Iroquois being armed. Those machines carrying troops had a machine gun mounted in each doorway to give covering fire to disembarking troops, while a specialist gunship version carried twin machine guns on outriggers on each side, with rocket launchers on racks below. Later, a grenade launcher was fitted beneath the nose.

The Iroquois was very successful in all its roles, but it was

too slow (at 138 mph top speed) to carry out the gunship role
as effectively as a faster machine could. What was needed was
a fast flying, fairly quiet aircraft, and a design competition was
announced to find a suitable helicopter. Bell were able to win
by very quickly redesigning the Iroquois to give the sleek, fast
Huey Cobra (AH-1G). This version had a tandem cockpit and
stub wings, with a top speed of 188 mph. The Huey Cobra, in
service in the late 1960s and early 1970s, was a formidable
machine with twin Mini-guns–six-barrelled, Gatling-type
weapons firing at a rate of up to 4,000 rounds a minute–76
2·75 in rockets or a 20 mm cannon. Experience with the Huey
Cobra has led to a new generation of gunships, including the
Lockheed Cheyenne and Sikorsky Blackhawk, designed from
the start as gunships and with speeds well over 200 mph. There
is no doubt that the Vietnam war, unhappy as it was in
international and diplomatic terms, was the conflict which
finally established the helicopter as an absolutely indispensible
battlefield weapon.

Among other major American helicopter producers were
Hughes, Boeing-Vertol, and Kaman. A notable Hughes machine
was the OH-6A Cayuse, a light observation type for the US
Army, over 3,000 being ordered. It is also widely used
commercially. Kaman's best-known design was the HH-43

above
The American McDonnell XV–1
was another VTOL design,
intended for military service

below
A Bell Iroquois gunship
launches rockets at enemy
positions in the Vietnam jungle,
in 1967

Huskie rescue and fire-fighting helicopter, a twin-boom machine
with foam equipment. With this, a technique of hovering above
any crashed aircraft to smother it in foam, then dropping rescue
personnel to the ground alongside the wreck, was developed.
It was standard USAF equipment for years, virtually a flying
fire engine.

Vertol was a company started by a famous helicopter pioneer,
Frank Piasecki, in 1945, later absorbed by Boeing. Of many
famous Piasecki designs, all of twin-rotor type, the most famous
is the Model 107, H-46, or Sea Knight. It has been used by
several air forces and commercial airlines in various forms,
carrying up to 25 passengers.

Helicopter development in Britain was much less prolific than
in America after 1945. Early attempts to produce entirely native
British designs included the Saunders-Roe Skeeter (similar in
many ways to the Bell 47), the Bristol Sycamore, the first British
designed helicopter to fly in 1947, and the Bristol 173 Belvedere,
a large, twin-rotor machine designed as a troop carrier. This
was used by the RAF though naval and civil versions were
contemplated. With the exception of the Sycamore, which
enjoyed limited export success (including Belgium and
Australia), these early British designs suffered mainly from a
limited market and could not enjoy the vast sales available to
the American types. The most successful helicopter firm in
Britain was the Westland Aircraft Company, whose fortunes
were based on building the best Sikorsky designs under licence
for the British and Commonwealth market. Westland did well,
and the changes made to the basic Sikorsky designs were often
genuine improvements. Westland fitted gas turbine engines, for
example, before Sikorsky did it themselves. In the 1960s the
only native design of any great significance was the Westland
Scout, a light utility army helicopter (also in service as a gunship
with machine guns and anti-tank missiles) and its naval
equivalent, the Wasp, which was an anti-submarine helicopter
for frigates. A more recent Westland success has been the Lynx,
an Anglo-French design which has produced an efficient utility
helicopter of similar style and performance to the Bell Iroquois.
Also produced in the 1970s by Westland, jointly with the French
firm of Aerospatiale, has been the Puma assault helicopter,
similar in size and style to the S-61, and the Gazelle, a light
liaison helicopter, to replace the Bell 47 (Sioux) type of machine

right
A Westland Scout, fitted with
SS–11 anti-tank missiles,
'hedge hopping' under cover

below
A Westland-Aérospatiale Puma
of the Royal Air Force lands
troops during exercises in 1973

centre, right
The Fairey Rotodyne of 1960
was a British attempt to produce
a VTOL machine – virtually half
helicopter, half conventional
airliner. It was later abandoned

in military service. It is interesting to note that the Italian
firm of Agusta has also thrived by building American types under
licence.

France has been particularly successful and forward looking
in helicopter design. Sud Aviation (now Aerospatiale) were
energetic in producing a prototype in 1948. From experience
with this, the Alouette series (Mks I, II and III) have been built
as general utility types, both for military and commercial
operators. The Alouette II, sold in more than 30 countries, was
particularly successful. It had the same sort of 'skeleton' tail
and Plexiglass nose as was used on the Bell 47. It had a rotor
diameter of $32\frac{3}{4}$ ft and length of $31\frac{3}{4}$ ft. A Turbomeca Artouste
turbine of 360 hp gave a cruising speed of 106 mph. The Alouette
could be used for casualty evacuation or could be fitted with
guns. It was supplied to police forces and several armies and
over 1,000 were built. The Alouette III had a fully covered tail
and a larger fuselage. Its 550 hp Artouste turbine gave a cruise
speed of 118 mph, and seven people (including the pilot) could
be carried. An armed gunship version was also produced.

In a larger category altogether was the Sud-Aviation Frelon
and Super Frelon, the latter being the production model. It was
produced in both cargo and anti-submarine forms. Some
technical liaison with Sikorsky facilitated its development. First
flown in 1962, the Super Frelon held several speed records in
its early days. Military versions of the Super Frelon have been
sold to South Africa and Israel. The cargo version carried up
to 30 troops, 18 stretchers, or up to 9,000 lb of freight. Smallest
of the Sud-Aviation helicopters was the Djinn, a liaison and
observation machine like the Bell 47. A feature of this helicopter
was the use of rotor-tip jets, one of its designers having worked
on the German Doblhoff 342 prototype previously mentioned.

In the Soviet Union some very quick developments were
achieved. Until 1950 there was little of note, though a design
bureau had worked on various helicopter and gyroplane designs
since 1940. The first design to emerge, the Omega, ran to three
prototypes, but it was not put into production. The inspiration
for the Omega, built between 1941 and 1944, was the Focke
Fw 61 which had impressed Bratakhin, the Omega designer,
and the machine quite closely resembled the latter in
appearance.

The Mi–12 was the world's
largest helicopter when first
flown in 1968. It has a
twin-deck interior. Rotor
diameter is 115 ft and fuselage
length, over 121 ft. An American
Bell Jet Ranger (length 31 ft)
is dwarfed in the foreground

In 1948 the first post-war helicopter design appeared, the Mi-1
(designer Mikhail Mil), which was very similar in appearance
and layout to the Sikorsky S-51, though was actually a little
shorter. The S-51 clearly inspired the design, though it was not
a direct copy. Known in the West as the Hare, it has an Ivenchko
575 hp radial engine giving a cruise speed of 87 mph and a
top speed of 106 mph. Thousands were built in both Russia and
under licence in Poland and served all Warsaw Pact countries,
plus other nations supplied by Russia. A turbine engine
development with enlarged cabin was designated Mi-2. The Hare
entered service in 1951 and was still in use in the 1970s.

Also from Mil came the Mi-4, known as the Hound in the
West. It was similar in layout to the S-55, though a little larger.
Clamshell doors in the rear fuselage, however, were a feature
not found in the Sikorsky design. Well over 3,000 were built
in Russia and China, and the type was standard in all air forces
under the Russian sphere of influence. There was also a
10-passenger civil version.

The first big Soviet helicopter was the Yak-24, a twin-rotor
type which could carry small guns and field cars, or up to 40
troops. This was not an entirely successful type. The Mi-6 of
1957, however, was quite sensational. It was the biggest
helicopter in the world at the time and the largest ever to go
into production. It was $108\frac{3}{4}$ ft long, $32\frac{1}{2}$ ft high, and had a
rotor diameter of $114\frac{1}{2}$ ft. It had a five-man crew and could carry
up to 65 passengers or 41 stretchers. Two Soloviev turbine
engines of 5,500 hp gave a top speed of 186 mph. Cars or light
guns could be loaded through clam doors at the rear. The Mi-6
had stub wings to assist forward flight. While officially a civil
helicopter, some saw military service. A further development,
and even bigger, was the Mil Mi-10, based on the Mi-6 but
designed as a flying crane. It had a quadruped undercarriage
and could pick up containers, pallets, or individual loads such
as heavy trucks. A lift of over 26,000 lb was possible, cruising
at 112 mph. The Mi-10 serves both the Soviet Army and the
national airline Aeroflot.

Though late into the helicopter field, the Soviet Union enjoys
a huge market in both Russia and the Warsaw Pact states. In
only a few years Russian designers have produced some good,
robust machines.

The Age of Air Power

The 'war to end wars' ended in November 1918 and the air forces which had been rapidly built up in the previous four years were just as rapidly run down. For several years after the war–indeed, for more than a decade in some cases–the air forces of the world largely made do with some of the aircraft with which they had ended the war in 1918. Thus the newly formed Royal Air Force used Vimy bombers, Bristol Fighters, and D.H.9As, among others, well into the 1920s. The overseas role of the RAF kept it busy policing the Empire air routes against dissidents in such areas as Iraq and the North West Frontier of India. Two-seater aircraft like the 'Brisfit' and D.H.9A were ideal for this sort of patrol work. Because of the overseas routes, however, the RAF did develop some troop carriers from the Vimy, the Vernon and Victoria which had large roomy fuselages similar to the Vimy Commercial, which saw service on some early British airline work. Thus the idea of moving troops by air was quite an early one, and useful experience had been gained by the time the Second World War broke out in 1939–when it did some of the Victorias of the early 1920s still served the RAF in Egypt!

New fighters and bombers were in the pipeline, however, although priority was not high. The 'first generation' of peacetime RAF aircraft were little more than refined replacements of First World War types, more powerful but aside from the engine showing no great technical advance over their predecessors. The Armstrong Whitworth Siskin and Gloster Grebe were the first new fighters. The Siskin had an all-metal airframe and a 425 hp Armstrong Siddeley Jaguar radial engine which was the first aero engine with supercharger to enhance performance at high altitude where the air is thin. The Siskin entered service in 1926 and at one time served with ten of the twelve fighter squadrons maintained by the RAF in the late 1920s (some Siskins were still in service in 1937). The top speed was 156 mph and the machine was armed with two .303 inch machine guns 'buried' in the engine cowling and synchronised to fire through the airscrew arc. Sighting at this time was by a telescope sight mounted by the pilot's windscreen.

The Vickers Vimy bomber was replaced by the Virginia, which was simply an enlarged version of the Vimy, a large ungainly biplane. As with airliners, so with fighting planes–the British authorities were conservative in the matter of design. So fabric covered biplanes were the rule until the mid-1930s when a belated rearmament began. There were numerous fine aircraft of their type, however, and if the aircraft tended to be outmoded by comparison with some other nations, the flying skill and training

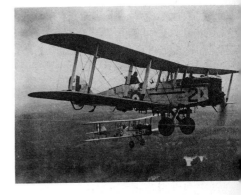

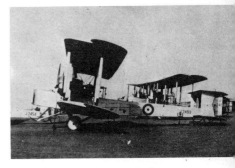

above
Vickers Vimys, 1920

top
D.H.9As of the RAF in 1926

opposite
RAF Bristol Fighters in their 'air control' role on India's North West Frontier during the 1920s

left
Bristol Bulldogs in 1932

above
A Hawker Fury in 1936

above, left
A Gloster Gladiator

of Britain's small air force was considered to be second to none.

Among famous RAF fighters of the 1920s, following the Siskin and Grebe, were the Bristol Bulldog, the Hawker Fury, Gloster Gauntlet, and finally the Gloster Gladiator, which was still in service in 1939 and served in diminishing numbers here and there until 1942, the last RAF biplane fighter. Of these types the Fury was one of the most aesthetically pleasing of all the biplane fighters. Sleek and streamlined, it was built round a 525 hp Rolls-Royce Kestrel in-line water-cooled engine, giving a top speed of 214 mph. The wingspan was 30 ft and length 26 ft 8½ in. In 1936 a Fury II was produced, with an uprated 640 hp Kestrel engine and a spatted undercarriage. Furies were famous for their polished aerobatic feats at the RAF Display at Hendon. They were in service until 1939. The Gauntlet replaced the Bulldog in 1935. With a Bristol Mercury radial engine it had a top speed of 230 mph. The Gladiator was a modified replacement introducing an enclosed cockpit and four .303 Browning machine guns, plus an uprated Mercury engine. In Norway, Greece, Malta and the Western Desert, as well as France in 1939–40, these quite obsolescent biplanes fought gallantly, though outmoded by German and Italian fighters.

The bomber situation in the RAF of the early 1930s was if anything worse. Money was tight, and the Air Ministry chose a light bomber ideal for colonial 'police' work, but again little more than a slight updating of World War I concepts. This was the elegant Hawker Hart, a two seater with the Kestrel engine and a top speed of 184 mph. Ordered in 1928, it was in service in 1930 and served until 1939. It carried 500 lb of bombs in racks beneath the wings. The Hart spawned numerous similar variants, including the Demon turret fighter with a hydraulically-operated turret, the Audax army co-operation machine, the Hardy for desert operations, the Hind, with an uprated engine, plus a trainer version of the Hart, the Nimrod fleet fighter, and the Hector. These fairly prolific, but inadequate aircraft made up the bulk of Britain's bomber strength. Heavy bombers came only in small numbers. The Handley Page Hyderabad and Hinaidi were descended from the 0/400 of First World War fame and were biplanes of similar conception. The Heyford replaced these (and Virginias) and was yet another biplane, which served from 1932–39. The one medium bomber squadron was formed in 1928 with the Boulton Paul Sidestrand, an open cockpit biplane which had two Bristol Jupiter 460 hp radials, giving a top speed of 140 mph. The bomb load was 1,050 lb. Its replacement was the Overstrand, a similar machine but with an enclosed cockpit and front turret, and bomb load increased to 1,600 lb. The Overstrand came into service in 1935; and as late as 1937 nearly every front line aircraft in the RAF was a biplane; such was the lack of real technological progress in the RAF until rearmament made some sweeping changes.

In the United States things developed slightly differently. In the fighter field, the splendidly graceful Curtiss Racers did much to improve the breed. Curtiss themselves developed a 'private venture' fighter from the Racers which took first two places for the US Army Air Corps in the 1922 Pulitzer race. This excited official interest and some prototypes were ordered, under the designation PW-8. In one of these, in June 1924, an Army pilot made the first US coast-to-coast flight in daylight hours to demonstrate the aircraft's speed. There followed a series of production machines, all variants on the basic design, as the Curtiss Hawk. The designation P (for 'pursuit') was chosen for this new generation of fast fighters. The P-1 Hawk was in service in 1926, and had a 435 hp Curtiss U-1150 engine giving a top speed of 160 mph; armament was two .30 inch machine guns. Hawks served well into the 1930s. The P-6E of 1931 was fitted with a 600 hp Curtiss engine and its top speed was up to 198 mph. The US Navy also had Curtiss Hawks under the designation F6-C.

The other major fighter builder in USA was Boeing. The firm got into fighters in 1921 by winning a contract to build 200 MB-3A fighters designed by Thomas-Morse. The MB-3A was one

of the fighters originated to equip the US Army Air Corps in 1918. Some native American designs were sought, and the MB-3A was based quite closely on the successful French Spad XIII fighter which equipped the American fighter squadrons. The war was over by the time the MB-3A was designed but production went ahead. This work inspired Boeing to design a better fighter in 1923 which was in turn influenced by the success of the Curtiss Racers. As the PW-9, a number of prototypes were supplied to the US Army. The engine and performance was comparable to the Curtiss Hawk – indeed the aircraft had a Curtiss in-line engine. Experience with the PW-9 led Boeing to build another 'private venture' prototype in 1928. The US Navy ordered this (as the F4B-1) and the US Army subsequently ordered the type as the P-12 as a Curtiss Hawk replacement. This neat biplane

top
The last of the Italian biplane fighters, the Fiat CR.42 of the late 1930s, which saw service in the Second World War

above
The Boeing-Stearman Kaydet was the principal US primary military trainer of the Second World War. This is a preserved example

Vickers Valentias flying over Egypt in 1935

had an aluminium tube framed fuselage, fabric covered wood-framed wings and metal-covered tail and control surfaces. When 90 were ordered by the US Army in 1929 it was the biggest order for an aircraft since the armament run-down had started in 1921. The new Pratt and Whitney Wasp radial engine had been built into the design, and in the final production form, P-12E, top speed was 189 mph. This model had a monocoque fuselage of new design.

Boeing's experience with all-metal aircraft–which included the Monomail mailplane–now led to a big step forward. At the end of 1931 they suggested an all-metal monoplane fighter and the US Army ordered three prototypes. They were delivered in 1932, underwent trials, and with some changes were ordered into production as the P-26A, with the first machines in service late in 1933. The P-26 was the famous 'Peashooter', the first monoplane fighter in squadron service. Top speed was 234 mph and the engine was a radial Pratt and Whitney Wasp of 500 hp.

The US Army Air Corps had a very small peacetime bomber force. From the MB-1 of 1918 was developed the MB-2, built in small numbers, essentially a biplane of the First World War era. One of these machines made history in July 1921 when General Billy Mitchell bombed and sank the ex-German battleship *Ostfriesland* during summer manoeuvres. He succeeded in demonstrating conclusively the vulnerability of big ships to air attack, though incurred displeasure among senior officers at the time. MB-2s served until 1928, and more biplane bombers replaced them, among them the Curtiss B-2 Condor and Keystone B-6 which were twin-engined types. The B-6 had Pratt and Whitney Wasp engines of 575 hp each and a top speed of 121 mph. Some 2,500 lb of bombs could be carried.

Monoplane bombers appeared in 1932. Boeing built the B-9, virtually a twin-engined version of the Monomail all-metal mailplane. Aside from prototypes, they did not win a production order, but the design became the basis of the Boeing 247 airliner which started a new era of flight.

above
The prototype Supermarine
Spitfire in 1936

top
Hurricanes of the American
'Eagle' Squadron in 1940

The successful contender for Army orders was the Martin B-10, an all-metal twin engine monoplane with retracting undercarriage and three cockpits. After trials a power-operated nose turret and cockpit covers were added. Twin Wright Cyclone radials of 675 hp each gave the B-10 a top speed of 207 mph faster than the Hawk and P-12 fighters then in service! The bomb load was 2,260 lb. Squadron service started in 1934 and B-10s in various forms served the US Army until 1939. There were export sales too, to the Dutch East Indies and these Dutch machines saw action against the Japanese in the Second World War.

Thus by the mid-1930s the US Army Air Corps had a very modern front line aircraft force–all monoplanes, all metal, and short only in numbers. America had taken the technological lead in warplanes, as in airliners and light aircraft, and was never again to lose it. Compared with the British, the secret of success was probably that the military authorities in USA were prepared to listen to the ideas and suggestions of the plane makers–all the successful US warplanes of the day had started as 'private venture' designs; in Britain the Air Ministry laid down very detailed requirements and asked for designs to suit. The problem with this approach was that the official requirements were all too often rooted in the past, so that the RAF was lumbered with *new* biplane fighter designs, for instance, right up to the outbreak of war in 1939.

Of the other major powers, France built few distinguished designs before 1935. One advanced type, however, was the Deiwoitine D.26 parasol monoplane of all-metal construction which was powered by a 250 hp Wright-Hispano radial engine, giving a top speed of 149 mph. The D.26 showed a different philosophy from both Britain and USA. The aircraft was tough and some were still flying in everyday use (as glider tugs) in the late 1950s. The D.27 of 1927 was a similar type, but with a 500 hp engine giving a top speed of 180 mph. When new this aircraft held the World Air Speed Record at 177·7 mph. The D.26 was also built in Switzerland for the Swiss Air Force. The

above
The Boeing P–26 Peashooter,
the first US monoplane fighter

opposite, top
The Bücker Jungmeister was a
prewar German sports aeroplane
used by the Air Sports
Organisation for training. This is
a postwar British civilian
example. The type was put back
into production for aerobatic use
after the Second World War

opposite, lower
A Messerschmitt 109E of 1940

Nieuport 62-C1 was a similar type of parasol wing monoplane, with a 500 hp Lorraine Petrel in-line engine. Later came a distinguished low wing all-metal monoplane, the Deiwoitine D.510, a clean fast machine with in-line engine. Italy also produced some fine warplanes, the Fiat CR.32 and Savoia-Marchetti SM.79 tri-motor bomber being the outstanding designs.

In the last half of the 1930s the entire picture changed. The leisurely days of periodically producing and introducing new types gave way to the frantic dash of rearmament, its pace only varying according to the country, and its political commitments. This great period of rearmament coincided with the new advanced aviation technology of the day, so the new aircraft which appeared made their predecessors look primitive and crude. It started with the rise to absolute power in 1933 of Hitler and the Nazi Party. By 1935 Hitler had rescinded the terms of the Versailles Treaty and new German armed forces were officially established and equipped—including an air force, the Luftwaffe. A very efficient biplane primary trainer, the Bücker Jungmann, and an advanced trainer, the Jungmeister, formed a sound basis of equipment for pilot training. Another classic trainer aircraft of the time in Germany was the Focke Wulf Fw 34. These trainers and others served with the Reich air sports organisation (Reichsluftsportverband), along with the gliders, as well as equipping the new Luftwaffe. The earliest Luftwaffe fighters were biplanes, the most important being the graceful Heinkel He 51, a beautifully streamlined machine reminiscent of the Curtiss Hawk and Hawker Fury. The He 51 had a 750 hp BMW V12 liquid-cooled engine and a top speed of 205 mph. Two machine guns in the engine cowling constituted the armament.

A significant development for the Luftwaffe was the move to a monoplane fighter. In 1933 the new German Air Ministry asked Bayerrische Flugzeugwerke to produce a single-seat

The Supermarine S.6B
Schneider Trophy winner, a
forerunner of the Spitfire

monoplane fighter following the trend indicated by Boeing with the new P-26. Professor Willy Messerschmitt was the designer of the new aircraft, and ironically enough the prototype, the Bf 109, had a British Rolls-Royce Kestrel engine since the German-made Jumo engine planned for it was not ready in time. In the design Messerschmitt disregarded official views, producing what he considered to be the essential form for the fast fighter of the future. The early machines built, the Bf 109A, B and C, were really development types appearing in limited service with the growing Luftwaffe in the 1936–38 period. In this period the Spanish Civil War was at its height and Germany supported Franco's Nationalists with the Condor Legion, recruited as volunteers from Luftwaffe personnel for service in Spain. The Civil War was an excellent proving ground for the new Luftwaffe–it gave fine experience to the men and helped iron out the snags with the aircraft as well as allowing tactics to be developed. With a 700 hp Jumo V12 engine, the Bf 109B spanned $30\frac{1}{4}$ ft, was $25\frac{1}{2}$ ft long and had a top speed of 292 mph. With a retractable undercarriage, all-metal construction, slatted wing edge for manoeuvrability, and beautiful aerodynamic shape, the Bf 109 (later Me 109) was an undoubted winner. So it proved to be, for in its later forms (Me 109E, F, G) it became the Luftwaffe's prime fighter and fighter-bomber, built in greater quantities than any other Second World War fighter (a vast group of companies produced over 15,000 of all variants). In terms of longevity–Spanish licence-built versions were still in service in 1965–the Me 109 lays claim to being among the all-time great aircraft. A specially prepared prototype version hoisted the world speed record to 379·38 mph in November 1937, incidentally.

As things turned out, it was the designer's ideas rather than official views which affected Britain's new generation of fighters. The Hawker firm (which had grown out of the Sopwith concern) produced the Hurricane prior to an official requirement to a design by Sidney Camm, the firm's chief designer. It was

128

basically a fabric-covered monoplane with metal-covered front fuselage, following the well-proven constructional features of the Hawker biplanes of the day. After the prototype had flown in 1935, the Air Ministry took an interest and placed an order, but Hawkers were bold enough (and confident enough) to lay down a production line for 1,000 machines before this. The engine was the new Rolls-Royce Merlin of 1,030 hp which give a top speed of 322 mph. An early machine made a record London-Edinburgh flight at 415 mph to show the potential of the monoplane fighter. The first production Hurricane flew in October 1937 and by the time the Second World War started 500 Hurricanes were available–largely due to the foresight of the Hawker firm.

By contrast the other new British fighter, the Supermarine Spitfire was an all-metal machine. It, too, was a 'private venture' design, offered to the RAF and accepted retrospectively since German rearmament was making British rearmament a necessity. R. J. Mitchell, the Supermarine designer, gained his experience from his work in producing successful racing floatplanes–the S.5 and S.6–for the Schneider Trophy races, which the S.6B won outright for Great Britain in 1931. The Supermarine floatplanes were built for speed and showed the economy of line and graceful proportions which were later reflected in the Spitfire. The prototype Spitfire flew in March 1936, 300 machines were ordered 3 months later, and first production aircraft were delivered late in 1938. The Spitfire at a top speed of 362 mph, and with a span of 36 ft 10 in, was smaller than the Hurricane (span 40 ft). The two aircraft complemented each other well–the Hurricane was tough, rugged, and less sophisticated, but its fabric covering made repairs easy. The Spitfire was faster and more dainty. In the Battle of Britain in 1940 the Spitfires would go for the enemy fighters while the Hurricanes took on the bombers.

Bomber development in Europe also took on a new look. The key influence here was again the Boeing 247 airliner. The Germans followed the Boeing's size, style and layout quite closely to produce the twin-engined Heinkel III, ostensibly as an airliner, but it was mainly developed as a bomber and became the mainstay of the Luftwaffe bomber force by the late 1930s. It had a bomb load of over 3,300 lb. The He III in various forms

below, left
The Vickers Wellington

below, right
The Focke Fw 200C long-range bomber was used to harrass and shadow Allied convoys

A Soviet ANT–6 bomber in use for paratroop dropping, 1940

was another long-lived type which survived in Spanish service into the 1960s. The Dornier company also produced a twin-engined machine, the Do 17, which was first advertised as a fast mail plane but was soon ordered into production as a bomber. Sleek and well streamlined the Dornier 17 was popularly called the 'Flying Pencil'. It carried 1,600 lb of bombs. Third of a trio of German bombers produced as prototypes in 1934 was the Junkers Ju 86, which was also designed as a passenger airliner but saw its main service as a bomber. Top speed was 224 mph and bomb load was 2,200 lb. This aircraft enjoyed export success and was supplied to several countries – Sweden, Chile, Portugal and South Africa.

In Britain a twin-engined metal monoplane, also on the lines of the Boeing 247, was designed by the Bristol Aeroplane Co. A prototype version, the Bristol 142, was built to the order of Lord Rothermere, owner of the *Daily Mail* and was named 'Britain First'. This flew in 1935 and achieved 307 mph, faster than any of the biplane RAF fighters then in service. The RAF immediately asked for it to be developed into a bomber and fighter; with only minor changes and the addition of armament it became the Bristol Blenheim. Deliveries began in 1937 and this very fast machine first saw service as a fighter. Two Bristol Mercury 730 hp engines gave it a top speed of 285 mph. Span was 56 ft 4 in and length 39¾ ft. A bomber version quickly followed and the Blenheim was in wide service in 1938–40, later serving in all theatres of operations. There were big export sales too. A Blenheim made the first British bombing sortie of the Second World War, just after war was declared.

The other distinguished British pre-war bomber was the Vickers Wellington. This was derived from the Vickers Wellesley, a single-engined long-range monoplane bomber, built as a 'private venture' in 1935. The RAF had ordered yet another 'new generation' biplane bomber, but Vickers decided to build the Wellesley to support their own view that a monoplane would be much superior. This proved to be the case. One of the designers was Barnes Wallis who had worked on the British Airship R.100. He used wood geodetic (lattice-like) construction which gave immense strength, similar to that used in the airship. The Wellesley was ordered for the RAF and in 1938 three of these machines flew from Cranwell, Lincs, to Ismailia, to set up a distance record of 4,300 miles. The Wellington was a twin-engined fabric-covered machine with geodetic construction and was in service by 1939. The geodetic construction was to prove extremely tough, and the aircraft could take immense battle damage and still fly home.

Among the other nations, the Soviet Union had some interesting designs. The big four-engined Ilya Mourometz biplane bomber had been flown in Russia before the First World War. The tradition was followed by a huge four-engined monoplane bomber, the Ant-6, which appeared as a prototype in 1930. It followed the Junkers method of construction with corrugated skin. Four 730 hp V12 engines gave a top speed of 122 mph and 4,850 lb of bombs could be carried both inside the bomb bay and on external racks. Span was $129\frac{1}{2}$ ft, and length 80 ft. For its day it was a huge and spectacular machine. Various models were built and among other duties it became one of the first paratroop carriers when Russia initiated experiments with the then novel idea of dropping assault troops into battle from the air during military exercises in 1937. During the Second World War the Russians used these big aircraft–they had over 800–for supply and paratroop work and made little attempt to use them for long-range bombing since they were slow and vulnerable by 1940.

Much more modern in concept was a twin-engined aircraft by the same designer, Tupolev. This machine, the SB-2, (or Ant-40), was another of the designs inspired by Boeing's B-9 and 247 aircraft. A twin-engined medium bomber, the prototype first flew in 1934, and machines were in service in 1936. Top speed was 255 mph, bomb load, 2,200 lb, and span was $66\frac{3}{4}$ ft. Over 6,000 were built. It was exported to China, Czechoslovakia and Bulgaria, and was supplied to the Republican forces in the Spanish Civil War (its German twin-engined equivalents all served with the Condor Legion on the Nationalist side). Two Russian fighters of the 1930s earned distinction, both with the Republicans in Spain and in the 1941–45 war against Germany. These were the I-15, a rugged manoeuvrable 229 mph biplane, and its monoplane derivative, the I-16, Rata ('Little Rat') which followed the Boeing P-26 Peashooter layout quite closely, but which was wood and fabric covered and had a retractable undercarriage. Both the I-15 and I-16 were in service to 1945 (and to the 1950s in Spain) despite their obsolescence.

below, left
Junkers Ju 52 troop carriers dropping paratroops, 1940

below, right
Yak–9 fighters of the Soviet-French Normandie-Nieman squadron in 1944. The Yak–9 was the most successful Soviet fighter of the war

By 1939 events had moved Europe close to war. On 1 September 1939, it was precipitated when Germany invaded Poland, and France and Great Britain, bound by treaty to Poland, declared war on Germany on 3 September. During the 1930s armchair strategists had prophesised that the next world war would be a cataclysm won from the air—capital cities would be wiped out by huge bomber fleets and civilians would die in millions. Such predictions were proved by events to be only partly true. Cities were raided, but no nation—certainly not Germany in 1939—ever had enough bombers to saturate a country with raids in the way so glibly predicted in the years before the war. Gas raids from the air never materialised either.

But if bombing attacks on cities were never to be as bad

as feared—although there were some massive and horrifying raids—the method of air warfare which did achieve quick results was in the tactical support of armies, which was perfected by the Germans in their spectacularly successful early campaigns of 1939, 1940 and 1941. Aircraft flew in ahead of the assault divisions and smashed key targets and defence points for the tanks and panzer grenadiers of the ground forces to occupy or overrun. The main instrument of this attack was the evil-looking Junkers Ju 87 'Stuka' dive-bomber. This machine had been designed in 1934, and was actually slow and vulnerable and poorly armed. But as a 'terror' weapon it had no peer in 1939 and did its job well—flinging a 500 kg bomb with some precision into its target area.

top
A Junkers Ju 87 Stuka in the Western Desert, 1941

above
The Junkers Ju 88, one of the best Luftwaffe machines of the Second World War. Originally designed as a bomber, it served in many roles—this is the Ju 88R–1 nightfighter version

The year of 1940 was an epic one for Western Europe. The first six months of the war after the invasion of Poland in September 1939 were quiet for Britain and France. Odd raiders over England were shot down and RAF Bomber Command made many sorties over Germany, dropping propaganda leaflets but not bombs—this was an altruistic period for British politicians. In April 1940 Germany invaded Norway; this was largely a sea assault, but key points in the south were taken by paratroops and soldiers flown in by Junkers Ju 52 to airports like Oslo and Kristiansand. On 10 May 1940, the German panzer armies rolled into France and Flanders with the awesome Junkers Ju 87 Stuka again softening up field targets for the assault troops. The RAF had an Advanced Air Striking Force in France, in the main equipped with Hurricane fighters and Battle light bombers. The Battle was a monoplane replacement for the Hart light bombers of the early 1930s. Though a handsome machine it was far too slow and hopelessly under-armed. Flying to attack German river crossing points on the Marne the Battles were shot from the sky by the German ground forces who had efficient AA guns organic at every level of command. The French Air Force suffered even worse butchery and the Luftwaffe had command of the sky, for the remaining British fighters were prudently pulled back to defend England—a wise move in view of what followed later in the year. British troops were drawn back to the port of Dunkirk by the end of May 1940, less than three weeks after the German offensive had begun, and the soldiers were taken back to England in the ad hoc fleet of 'little ships'—pleasure boats, fishing boats, and so on—hastily mustered for the occasion. Over the Dunkirk beaches British fighters kept the German air force at bay, flying from England. Another new fighter made a brief but spectacular appearance at Dunkirk—the Defiant, a monoplane which took a heavy toll of German fighters attacking it conventionally from astern and meeting a hail of bullets from four machine guns in a turret behind the pilot's cockpit. The measure of the Defiant was soon taken, however. It had no frontal armament and results in the air battle over south east England in the summer of 1940—the Battle of Britain—showed that the eight machine guns in the wings of the Spitfire and Hurricane were the most efficient of plane killers. The Me 109E of the day was cannon armed and pointed to the future. The new marks of Spitfire and Hurricane which were to enter service in 1941–42 were also cannon armed, giving an aircraft more stopping power as speeds got faster and armour plating was introduced around cockpits of the German machines.

The Defiant was switched to the night fighter role as the Germans, failing to gain mastery of the skies over England, switched to night bombing. A new word, the 'blitz' came into vogue for this sort of attack. London, Coventry, Southampton, Bristol and other centres of commerce and industry took a heavy pounding. The older Blenheim fighters also became night fighters and the Beaufighter, a faster replacement for the Blenheim, became the first of a new generation of successful radar-equipped night fighters which gave the pilot 'eyes' at night.

Radar now became a key weapon. Radio-locating, as it was first called in Britain, was developed with various degrees of success in Britain, Germany and America, as a spin-off from the research and development of radio and television which took place in the late 1920s and 1930s. Radar as a defence weapon was effectively developed in Britain by about 1936, by a team of scientists under Robert Watson-Watt. The salvation of Great Britain in 1940 owed as much to the ring of Chain High and Chain Low radar stations around the coastline as it did to the hard-pressed fighter pilots of the RAF. Radar detected the approach of raiders at long range so that fighters could be airborne and vectored to the enemy machines by the time they were over the English coast. It was efficient and an almost foolproof method of detection. It obviated the old wasteful system of rotating airborne patrols which had to be flown in the days of visual spotting. Radar also worked in all states of the weather, and around the clock. Soon aviation electronics became an industry in its own right, and radar has become an integral part of the aviation world, largly unseen by the layman (apart from prominent electronics aerials) but revolutionising air traffic control and making a major contribution to air safety as well as to defence. Its importance was greatly increased after the Second World War as aircraft speeds went up and the air space in the civilised world became more and more congested with the expansion of commercial airline flying. As the war progressed radar became an instrument of offense with the RAF and USAAF. Various navigation and bombing aids evolved, such as 'Gee' and 'H$_2$S' enabled bomber aircraft to be directed to their target and 'see' an electronic 'map' of the ground area below. Radar was miniaturised sufficiently early on to enable it to be fitted to aircraft used in the night fighter role–the twin-engined Beaufighter which entered service in 1941 was the first so fitted. The Beaufighter was a brilliant design, powerful with two 1,500 hp Bristol Hercules radial engines, and a steady platform for four 20 mm cannon and six machine guns. It was yet another 'private venture' of great merit and it was soon joined by an even more spectacular partner, the de Havilland Mosquito which became one of the most famous aircraft of the war.

A Curtiss P–40N in 1942

above, left
The de Havilland Mosquito,
shown in service in 1944

above, right
An F–51D Mustang of the
South African Air Force in Korea
in 1950

The Mosquito was another 'private venture'. It was built almost entirely of moulded plywood—hence its nickname 'The Wooden Wonder'. Two Rolls-Royce Merlin engines gave it a top speed of over 370 mph and the original de Havilland suggestion was that it be used as an unarmed light bomber, relying on speed for its defence. In the event it was used for almost every purpose—bomber, torpedo bomber, night fighter, fighter, photo-reconnaissance, and fast courier to 'run the gauntlet' with passengers for neutral Sweden. It was also used as a trainer, and made a special name for itself as a 'pathfinder'—the navigation lead aircraft which marked the targets with flares for heavy bombing raids. The Beaufighter was also used in several roles besides night fighting—it was a day fighter, anti-shipping bomber with rockets and torpedoes, ground attack fighter, and in post-war years a fast target tug. To the Japanese in the Far East where Beaufighters were used extensively by the Royal Australian Air Force, it was known as 'Whispering Death'.

When the RAF was hard pressed for aircraft in 1940–41 new aircraft began to arrive from America. The American plane makers had produced some very advanced types, some othem superior to anything seen in Europe. The Lockheed 14 airliner was developed into the very successful maritime patrol bomber, the Lockheed Hudson, for the RAF. The Hudson had all the refinements for which the Lockheed airliners had become famous. The Hudson was one of the first British purchases even before war was declared. It supplemented the slow but reliable Avro Anson which Britain used for maritime patrol as well as training. Another early purchase was the superb North American Harvard (T-6 or Texan in USA), the classic training machine with the distinctive 'buzz-saw' engine note which trained many thousands of pilots, eventually served with scores of air forces and was still in service in the 1970s, having first been produced in 1938.

The French had ordered a fine new Douglas light bomber, the DB-7, in 1938. This was a streamlined mid-wing machine with two Pratt and Whitney Double Wasp radial engines of 1,600 hp giving a top speed of 347 mph. It carried 2,600 lb of bombs. Deliveries had barely started in 1940 when France was overrun, and the aircraft were diverted to the RAF. This led to further orders and the aircraft, known variously as the Boston (RAF) and Havoc (RAF and USAAF) and A-20 (USAAF) was one of the most famous and familiar Allied types of the war. It served in many roles—fighter, night fighter and light bomber, in various configurations.

France had also ordered one of the first of America's new modern fighters, the P-36, Curtiss Hawk. The US Army Air Corps held a competition in 1935 to choose a new fighter and Curtiss and Seversky (with the P-35) submitted designs, both of which were ordered in small numbers. The P-36 had a Wright R-1830 1,050 hp engine giving a top speed of just over 300 mph. Armament was two machine guns in the engine cowling. Hawks were sold widely overseas to France, Finland, Peru. Some ordered for Norway were diverted to Canada when Norway was occupied and others for France went to the RAF. They had been outmoded by 1940 and were little used by the RAF, but a development of the P-36, the P-40 came in 1939. Similar to the P-36, the P-40 had an Allison liquid-cooled engine and guns in the wings. Top speed was 350-380 mph depending on model. From the P-40A to P-40N there were many models from 1940 to 1945 and the P-40 was the third most widely produced US fighter (to the RAF it was known as the Kittyhawk, Tomahawk, and Warhawk depending on model). The later models had an American-built Rolls-Royce Merlin engine which gave a superior performance. The RAF and USAAF used these aircraft extensively in the fighting in the Western Desert. The heavy radiator 'jowl' of the P-40 lent itself to sharkmouth decoration which was a popular motif for these machines. Most famous of the P-40s were those used by the American volunteer pilots, the 'Flying Tigers', fighting for the Chinese in the 1940–42 period.

above
The P–47D Thunderbolt, the biggest single-seater fighter of the Second World War

top, left
First service version of the North American Mustang, the P–51A of 1942

top, right
A P–38F Lockheed Lightning in 1943

A Beaufighter at Malta used for anti-shipping strikes, June 1943

The greatest American fighters were yet to come, however. Among several of distinction the P-51 Mustang, P-47 Thunderbolt, and P-38 Lightning were the most famous. The Mustang was built to meet British requirements, since the RAF found previous American types unsuitable in various ways. North American, the makers, designed it under a 120 day contract period. The prototype flew in October 1940 and had a British 'look' about it – especially as an in-line engine was fitted and the fuselage had sleek contours. It was technologically very advanced, with radiator set well back for streamlining. The US Army Air Corps was impressed and ordered the aircraft too. A dive bomber variant was the A-36 and there was also a photo-recce version. The early models were not entirely successful, but the Merlin engine was adopted to replace the Allison unit and there was a dramatic improvement in performance. The P-51C and P-51D (which had a 'teardrop' cockpit cover for enhanced pilot visibility), were the improved models. Top speed of the P-51D was about 450 mph, while a lightened version, the P-51H had a speed of 487 mph. The Mustang was very widely used by all the Western Allies, and the type was operational in the Korean War in the 1950s. In 1967 limited manufacture started again by the Cavalier Aircraft Corp, to provide a suitable aircraft for counter-insurgency (COIN) operations, making the Mustang a fighting plane of remarkable longevity. In post-war years Mustangs served many air forces, and also became popular as racing and fast executive aircraft.

The P-47 was developed from the Seversky P-35 and the Republic Lancer, P-43. The P-47 Thunderbolt designers greatly enlarged on previous types. They built the plane round a big 2,000 hp Pratt and Whitney R-2800 radial engine and went for range, speed and toughness. This resulted in a big fighter for its day – 40¾ ft span and 36 ft length, empty weight 10,000 lb. It was the biggest single-seater fighter produced in America. Its immense size and strength earned it the nickname 'Jug' (for Juggernaut) and it could take terrific punishment.

The Lockheed Lightning (P-38) was distinctive in having a twin-boom fuselage layout. With two engines in the booms, the pilot was housed in a central nacelle which carried a heavy cannon and machine gun armament in the nose. Early models had some 'teething' troubles, but the type was subsequently a great success.

All these American fighters had the great merit of a good range – lacking in the small British fighters. Thus when the bombing war was carried to the German homeland from 1942 onwards the availability of the P-47, P-51 and P-38 ensured air cover for the bombers and it is significant that before the escort fighters became available the American day bombers suffered heavy losses at the hands of German fighters. The American fighters were also a good counter to the Focke-Wulf Fw 190, the superb new 'second generation' radial-engined fast fighter which the Germans developed.

The big day bombers which the Americans developed were the other major contribution to the war effort. The Boeing company started work on a four-engine bomber in 1934 to meet a US Army requirement for a bomber with a 2,000 lb payload, a range of 1,020 miles, and a speed of 200–250 mph. Heavy armament was specified and several gun positions were provided

A Focke Wulf Fw 190D, the
Luftwaffe's finest
piston-engined fighter

in the design. This led to the name Flying Fortress. Prototypes of the design were tested in the 1935–37 period, and early models were in service in 1938–39. These showed up several design faults. A larger tail was needed to improve stability and the chance was taken to add a tail turret, and improve the fire positions in the fuselage. Resulting models, the B-17E, F and G, all variously modified, saw service in Europe after America entered the war against Germany. The B-17 formed the backbone of the 8th Army Air Force which was based in England in 1942–45. The first big daylight bombing raid (on Rouen) took place in July 1942. The first big attack on Germany took place one year later when 376 B-17s attacked the important ball bearing factory at Schweinfurt. Despite the heavy armament, 60 B-17s were lost, and a second raid led to similar losses and only the coming of long-range escort fighters saved the day. Some 12,677 B-17s were built in all and though most were used as strategic bombers, some were handed to the RAF as maritime patrol aircraft, and others were used as cargo planes and after the war as rescue aircraft by the USAF. The B-17G had a top speed of 287 mph and a range of 2,000 miles with 6,000 lb of bombs.

The other great American bomber was the Consolidated Liberator, B-24, which was designed in 1939 to supplement the B-17. A tricycle undercarriage and high wing were new characteristics, and up to 8,800 lb of bombs could be carried in a capacious bomb bay. Top speed was 300 mph from four 1,200 Pratt and Whitney engines. Nose, tail, ventral, upper, and side gun positions were provided. Span was 110 ft and length 67 ft. The first B-24s were in service in 1941. War now seemed inevitable and huge production orders were placed. A big syndicate of constructors was involved and the Liberator became the most extensively produced American bomber. Models from B-24A to B-24M were produced. Some were supplied to the RAF for maritime patrol work where they did good work against German U-boats. The long range of the Liberator, made it ideal for ocean patrol work and it was now possible to cover the Atlantic with aircraft; previously there had been a mid-ocean 'gap' where U-boats were immune from air attack. Most Liberators were used in the bombing role, however, in Europe, the Middle East, and Far East—huge raids by Liberators were carried out on Japanese targets in the Pacific theatre.

Many other aircraft won fame in the Second World War, but only a few of these will be remembered as truly 'great' aircraft long after most others are forgotten. The Lancaster bomber achieved immortality for its outstanding operational record with the Royal Air Force and other Allied air forces working under RAF command during the war years; it was the spearhead of RAF Bomber Command's huge bombing offensive against Germany and was the principal RAF heavy bomber in the second half of the war; it was specially adapted for hazardous raids against key targets, such as the Ruhr dams; it saw much service both during and after the war as a maritime reconnaissance aircraft; in post-war years it was developed as a 'stop gap' civil aircraft and remained in RAF service well into the 1950s and with other air forces even longer. Over 7,000 Lancasters of all types were built, and the 'Lanc'–as it was popularly known–was truly an example of the 'right aircraft at the right time', which was the key to its great success in service.

The story of the Lancaster goes back to the mid-1930s when the British Air Ministry began a modest expansion plan to ensure that the RAF had 500 bombers by 1935. By the end of 1935, however, the Italian conquest of Abyssinia and the rearmament of Nazi Germany under Hitler's leadership, led to the expansion target being doubled to 1,000 aircraft with over 800 being required by 1937. In May 1936 formal specifications were put out by the Air Ministry for both twin-engined and four-engined types, and heavy bomb loads (by the standards of time) were called for, of between 8,000 and 12,000 lb, plus bomb bays big enough to carry torpedoes. Earlier specifications had already resulted in the development of simpler twin-engined types, the Wellington, Whitley, and Hampden, and these were destined to be the bombers which formed the main part of the

Consolidated Liberator was built in vast quantities–total production of all versions exceeded 19,000 aircraft. It was most familiar in its bomber and maritime reconnaissance versions; this aircraft is an unarmed transport in RAF markings, although it was operated by BOAC

A Boeing B–17G, the classic Flying Fortress, in 1944

The Douglas Boston III in RAF service in 1941

RAF's striking strength in the 1939–41 period. However, the 1936 specifications led to the development of the types which would ultimately succeed these earlier monoplane bombers, and of these the Stirling (built by Short Bros) was the first four-engine bomber to enter RAF service, in 1940. The Handley Page Halifax followed it at the end of 1940 and was used by Bomber Command until 1945 with great success, even though it was eclipsed by the Lancaster, both in the number of sorties flown and the weight of bombs dropped on the enemy. The Stirling had a more chequered career and was relegated to secondary roles once the Lancaster became available in large numbers.

Prior to the development of the Halifax, Handley Page had proposed a version with twin-engines (the HP 56)–which was subsequently dropped in favour of the four-engined Halifax–and A. V. Roe Ltd proposed a design known as the Avro 679. This was a large twin-engined machine powered by Rolls-Royce Vulture engines. The Vulture was a somewhat complex engine, essentially two V12 Kestrel engines on a common crankshaft, one above the other to give a ×24 cylinder layout. The original Kestrel engine from which the Vulture was evolved was well proven, having powered the Hawker fighters (and other types) in the early 1930s. As the Vulture power units were each 'twins', in effect, the Avro 679 was to all intents and purposes four-engined, even though it had a twin-engine layout. A wooden mock-up of the Avro 679 was built and some 200 machines were ordered before a prototype had flown. The name Manchester was given to the new aircraft, and the unarmed prototype made its first flight in July 1939.

While the Avro Manchester was structurally a sound aircraft, it was an almost complete failure in terms of suitability for service. Concentration on twin-engined types rather than a four-engined type from the start was influenced by opposition in some quarters to large four-engined bombers, due to the expense that would be incurred in extending runways and airfields to enable them to operate (defence budgets were extremely tight, even in the late 1930s when rearmament was under way). Some largely abortive experiments with catapult launching were put in hand to help overcome this problem while work proceeded with the Manchester. The weakness of the design was in the Vulture engines, which had been fairly hastily developed and ordered without any really satisfactory trials

programme, due to the urgent need for new bombers. The engines
failed to give their intended power output and this in turn made
the relatively heavy aircraft unstable; the wingspan was
extended by ten feet and an additional fin added on the fuselage
rear in an attempt to improve the stability. More than a year
was taken up in trying to overcome the design problems, and
the war had been under way for over a year before the first
Manchester was delivered to the RAF at the end of October
1940. No.207 Squadron took the early aircraft which were still
dogged by technical troubles, almost all caused by the defects
of the engines. The first Manchester operational sortie took
place against Brest dockyard in February 1941, but overall the
type had a limited career due to frequent engine failures which
caused aircraft to be suspended from operations for long periods
while attempts were made to remedy the problems.

Despite its ignoble service career, the Manchester was the
aircraft which made the great Lancaster possible. The
Manchester airframe was extremely rugged and sound, and when
it was realised that the Vulture engine troubles were likely
to persist, the idea of modifying the basic airframe to take four
Rolls-Royce Merlin engines was taken up. The Merlin was
already proven; it powered the Spitfire and Hurricane fighters
which had been built in large numbers, and the four-engined
layout had been proved a practical proposition in the Handley

Page Halifax which was already in production at the end of 1940. Thus a Manchester airframe was adapted from the production line and completed with four Merlin engines; it still retained the central tailfin and was, in fact, originally known as the Avro Manchester Mk III. The prototype flew in January 1941 and proved successful from the start, so successful that Manchester production was ordered to be converted to the four-engined configuration from the 200th machine onward.

All this accounted for the speed with which the Avro Lancaster was got into production and, almost immediately thereafter, into RAF service. Conceived as an afterthought, almost, the Avro Lancaster proved to be the best of all the bombers which served with the Royal Air Force in the Second World War. It was docile to fly, took terrific punishment both from the stresses of flying and from enemy action, and it was easy to maintain and very reliable mechanically. Above all, it had inherited a capacious and uncluttered bomb bay which not only allowed it to carry 4,000 lb bombs (in 1942 the largest in use) but subsequently enabled it to carry massive bombs up to 22,000 lb in weight as developed later in the war (by contrast the Stirling and Halifax, being developed earlier, had compartmented bomb bays which restricted the size of individual bombs which could be carried). Aside from engine changes and equipment changes, the Lancaster was virtually unaltered in basic design throughout its entire service life, a fairly rare thing with military aircraft.

The RAF bombing offensive was largely orientated to night attacks, though some daylight raids were undertaken. Through 1942 Manchester squadrons converted to Lancaster squadrons and by March 1943 there were 18 Lancaster squadrons with Bomber Command. By 1945 the Lancaster force had built up to around 50 squadrons. Most of the Lancasters used were Mk Is (or the externally-similar Mk III which had American-built Packard Merlin engines) and these represented the major production type. Some 608,612 tons of bombs were dropped by Lancasters by 1945, most of them on Germany or German held territory and 156,000 sorties were flown. These impressive statistics include the Lancaster's participation in the famous 'Thousand Bomber Raids', huge efforts which started in mid-1942 and involved bringing in all sorts of bomber aircraft (some from Coastal Command) to reach the 'magic' 1,000 total. In fact, Thousand Bomber Raids were not maintained at this level for long and to a great extent the increasing availability of the ever-reliable Lancaster made it possible to put up very effective raids over the Ruhr and Berlin with much smaller bomber forces.

The Lancaster was also used for the celebrated 'dams raids' on the Sorpe, Ede and Moehne dams in an attempt to flood the Ruhr area and disrupt industry. Dr Barnes Wallis produced an idea for a spinning bomb which would skim across the surface of the reservoir to the target dam, sink alongside it and split the dam with an 'earthquake' effect. Lancasters were specially modified to carry the cylindrical shaped bomb, with a car engine inside the aircraft to drive the bomb on a spindle to impart the spin before it dropped. The raid, which involved skilful and hazardous flying, took place on 16 May 1943; it was brilliantly carried out by a specially assigned squadron with fairly heavy losses, though the effect was not as conclusive as had been hoped.

opposite, top
The classic Fairey Swordfish torpedo bomber

opposite, lower
A Fairey Albacore, the later derivative of the Swordfish, is 'bombed up' for a sortie in 1943

The ultimate in heavy bombers by 1945 was the B-29 Super-fortress, successor to the B-17. At the time of its appearance in service at the end of 1943 it represented three years of intense development work. Boeing had projected some very long range bombers back in the 1930s, and one prototype with a 5,000 mile range, the XB-15, had actually been built. Further development led to a project which Boeing was asked to develop in 1940, for a bomber able to carry a 2,000 lb bomb load for 5,300 miles and with a 400 mph top speed. The aircraft was ordered straight into production late in 1941 without going through the usual prototype stage. Four 2,200 hp Wright R-3350 turbo-supercharged radial engines gave a top speed of 358 mph. For high altitude operations the aircraft was fully pressurised, in two sections, front and rear, joined by a sealed tunnel passing above the bomb bay. Sperry remote-control turrets covering all arcs of fire were a big defensive step forward. The major difference in the B-29 from the original requirement was the bomb load of 20,000 lb, a reflection of the way actual combat experience affected design requirements, for one of the major Second World War developments was in the size and types of bombs. By 1945 the two most important Allied bombers were those which could carry the biggest bombs, the Lancaster and the B-29.

The B-29 was deployed in Pacific area operations against Japan from bases in India, China and later in the Marianas and other Pacific islands. Through 1944 and 1945, B-29s hammered Japanese targets mercilessly, fire raids with incendiary bombs on Tokyo and other major cities being a feature of the campaign (the flimsy houses of Japanese urban areas burnt easily). Air opposition was so light in Japan by 1945 that B-29s had most of their defensive armament removed so that even bigger bomb loads could be carried. The B-29 was instrumental in the sudden end of the war. The ultimate weapon, the Atomic Bomb, had been developed and produced in America and on 6 August 1945, a B-29 named 'Enola Gay' dropped the first ever A-bomb used in warfare, the 9,700 lb 'Little Boy' at Hiroshima. A second B-29, 'Bockscar' dropped another A-bomb, 'Fat Boy', on Nagasaki, three days later. When the full horror of this type of attack struck home, the Japanese sued for peace and the Second World War was over.

Naval aviation made enormous progress in the Second World War. No pure aircraft carriers had been operated during the 1914–18 war, though there was a very active and largely land based Royal Naval Air Service. In Britain three large battle cruisers, *Furious*, *Courageous* and *Glorious*, were converted to give flying-off decks forward and landing-on decks aft of the superstructure and funnels. With this somewhat clumsy arrangement, intrepid pilots managed to land and take-off. Some capital ships had wood platforms on their turrets and such land plane types as Sopwith Pups and 1½ Strutters could take off, but there was no way of landing on so that operations had to be within range of land. Naval air operations were thus usually in the hands of seaplanes like the long-lived Short 184. The seaplane carrier became a new class of ship, lifting aircraft by crane onto the water for take-off and recovering them again after flights.

In 1918 the Royal Navy commissioned *Argus*, the first 'flat-top' aircraft carrier, a converted liner, which had the now familiar shape associated with this type of ship. The US Navy followed on in 1922, with *Langley*, a converted collier. France completed the battleship *Béarn* as an aircraft carrier in the late 1920s, by which time the Royal Navy had converted their three ex-battlecruisers into proper 'flat top' carriers. The US Navy converted two modern battlecruisers, *Saratoga* and *Lexington* into the world's largest carriers by the early 1930s. Japan with *Akagi* and a host of others began to build one of the largest carrier fleets. In the late 1930s purpose-designed aircraft carriers were coming into service, *Ark Royal* and others in the Royal Navy, and *Hornet*, *Enterprise* and others in the US Navy. These all had greatly increased aircraft capacities, and special techniques for aircraft operation had been evolved. Arrester wires, barriers, and catapults or accelerators were used to overcome the limitations of length of landing run and take-off run in the 450–700 ft length of typical carier flight decks. This in turn imposed necessary modifications to aircraft designs for carrier operation. Arrester hooks, folding wings for close stowage, and catapult attachment points became features of all naval aircraft.

A Consolidated Catalina flying boat refuels from a seaplane tender in the Aleutians, 1943

The Royal Navy's Fleet Air Arm was very much a 'Cinderella' service between the wars. The old RNAS had become part of the RAF in 1918 and the RAF administered the Fleet Air Arm until 1937. Largely as a result–and because defence funds were short–the Fleet Air Arm struggled along with obsolescent aircraft, and in 1939–40 was equipped with some distinctly inferior machines. Despite this the Fleet Air Arm did well, an early feat of arms being the very successful raid on the Italian base of Taranto in November 1940, when carrier-borne aircraft attacked and immobilised several major Italian warships. The instrument of this attack was one of the most famous of aircraft, the Fairey Swordfish biplane torpedo bomber. This aircraft looked venerable, and it was, for it dated from 1934. A 690 hp Bristol Pegasus engine gave it a top speed of about 135 mph. It had a three-man crew in open cockpits. Nonetheless, the Swordfish remained in production and in service throughout the war, even outliving its successor, the similar but more refined Albacore, as well as all its contemporaries. The Swordfish just happened to be a rugged and viceless aircraft, very adaptable. Later models had search radar and carried depth charges and rocket projectiles for the anti-submarine role.

The US Navy and Japanese air arms took more care in the 1930s to keep their equipment up to date. From the Curtiss Hawk in the mid-1920s there was a succession of up to date fighters and attack bombers, including such classic types as the Boeing F4B (equivalent to the Army's P-12), the Curtiss BFC, and the Grumman F3F. For a short time in the early 1930s, the US Navy also had two airships, *Macon* and *Akron* which were fitted to carry, launch, and recover four aircraft each. A special fighter, the Sparrowhawk, was developed for these airships, featuring a special recovery pylon on the upper wing. The idea was interesting but both the airships had crashed by 1935.

The US Navy was quick to get into the all-metal monoplane era. The Douglas Devastator was in service in 1937 to equip the new generation of carriers. The Wildcat fighter, Dauntless dive bomber, and Buffalo fighter (unsuccessful as a Navy fighter, but widely exported) were among the first generation of monoplanes. The Wildcat was supplied to the Royal Navy to make up for the British lack of modern fleet fighters in 1940. A second generation of fleet aircraft included the superb F6F Hellcat and F4U Corsair fighters, and the Helldiver and Avenger bombers. Of these the Hellcat, Corsair and Avenger also served with the Royal Navy and between them gave the Allies total dominance in the air war at sea against Japan.

The Japanese Navy air arm was something of an unknown quantity in the late 1930s, since Japan kept very much to herself. When hostilities started in late 1941 between Japan and USA and Britain, Japan administered a rude shock to the Allies. The Japanese Navy had early realised the importance of naval air power and her carrier fleet was well equipped. The Mitsubishi A5M fighter was a fine metal monoplane with 270 mph top speed, and its famous successor was the A6M Zero-Sen, a formidable, rugged, and efficient fighter, superior to the early American fleet fighters. Its 1,130 hp Nakajima radial engine gave a speed of over 350 mph and the armament was two 20 mm cannon and two machine guns, plus a 700 lb bomb load capacity.

In the Aichi D3A2 (Val) the Japanese had a tough dive bomber, and the Nakajima B5N and Yokosuka D4Y Suisei were other important bombing types. The raid on Pearl Harbor in December 1941 was an even bigger success than Britain's Taranto raid of the previous year. Unfortunately for the Japanese, however, the US carrier force was at sea on the day of the raid and the American carriers were left intact to carry the naval air war back to the Japanese fleet in four years of hard campaigns and epic battles. These included the Midway battle in which the opposing naval fleets did not actually meet – all the fighting was done by the aircraft of the two fleets.

During the Second World War there were substantial carrier building programmes in the USA and Britain, and fleet carriers became bigger to carry more aircraft. On the other hand the escort carrier, known as a 'Woolworth' or 'Baby Flat-top', was evolved, basically a merchant ship hull with a wooden flight deck and carrying about 20 aircraft. While intended initially to provide air cover and anti-submarine aircraft for convoy protection, escort carriers were also used to provide fighter or fighter-bomber support for amphibious assault operations, or to carry replacement aircraft to supply the attack carriers of task forces. The light carrier was another new type, a cheaper lighter version of the fleet or attack carrier, which was quicker and cheaper to build, but carried fewer aircraft. The Royal Navy and US Navy both introduced light carriers.

above, left
F6F Hellcat in service in 1943

above, right
An A6M Zero-Sen, the most famous Japanese fighter of the Second World War

below
Carrier strength against Japan: Corsairs prepare to raid Japanese positions in Sumatra from the British carrier *Illustrious* in September 1944

right
RAF Lightning intercepts a
Soviet Tupolev Tu–20 'Bear'
long range reconnaissance
bomber over the North Sea in
1974, in a typical 'cold war'
incident

below
An RAF Nimrod patrol
aircraft – developed from the
Comet airliner – flying over the
Soviet helicopter carrier
Leningrad in 1974

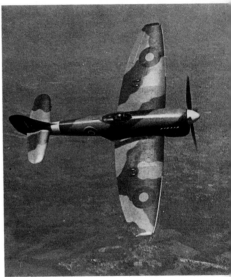

The availability of carrier aircraft led to the abandonment of catapult-launched seaplanes carried on capital ships and cruisers. By the end of the war, experiments were already being made in landing the new generation of jet fighters on carrier decks, and this was to open up yet another era of carrier operation from the 1950s onwards.

The latter half of the Second World War saw new aspects of air warfare being developed. The glider came into its own as a means of landing troops en masse and their heavy equipment in support of parachute operations. Parachute troop landings by the Germans in their 1940 campaigns, and by the Allies in the 1944–45 period, were in themselves manifestations of air power. Allied gliders included the Horsa and Waco Hadrian, and the massive Hamilcar which was built to hold a light tank. Tanks were flown into battle for the first time during the airbone drop in Normandy in June 1944. The glider was vulnerable, and clumsy, however, and had a short life. As soon as bigger and more capacious military transports were available in post-war years, and big troop-carrying helicopters were in prospect, the glider rapidly disappeared as a military weapon.

above, left
A Hawker Typhoon

above, right
A Hawker Tempest of 1944

top
Glider operations in the Second World War: a Waco Hadrian is hauled off by a C–60 transport aircraft

The tactical use of aircraft in support of ground forces was
effectively demonstrated by the Germans in their 1939–40
campaigns. As the Allies carried the war back to the Germans
and Japanese, they carried the idea further – as did the Russians.
Such aircraft as the Hawker Typhoon (the intended successor
to the Hurricane), the P-47 Thunderbolt, the P-39 Airacobra,
and the Soviet Il-2 'Stormovik' became legendary ground attack
machines with bombs and rocket projectiles, prowling the skies
in support of ground forces. In Normandy in 1944 more German
tanks were destroyed by rocket-firing aircraft than by Allied
anti-tank guns, so effective was Allied mastery of the sky at
that time.

Supply by air on the grand scale was also something new.
The German 6th Army at Stalingrad, and the Chindit raiders
behind Japanese lines in Burma were just two of many
operations sustained entirely by air while ground forces were
cut off from their own lines. The Berlin Airlift in 1948–49 was
an even more dramatic demonstration of the new capability of
air transport to shift cargoes effectively and quickly, given
suitable aircraft. These experiences in the 1940s led to the bigger
and bulkier designs of cargo aircraft evolved in the 1950s and
1960s.

Towards the end of the Second World War the jet propulsion
era became a reality and started to transform the style of air
warfare yet again. Several pioneers worked on aspects of jet
propulsion, years before it became a practical type of aircraft
power unit. The theory was explained by several scientists in
the nineteenth century but the wherewithal was lacking. An

top, left
The first British jet aircraft, the
Gloster E.28/39, of 1941

top, right
A Messerschmitt 262A, the first
Luftwaffe operational jet aircraft

above, centre
A de Havilland Vampire III of
1946

above
The Soviet MiG–15 of Korean
War fame

English scientist at the Royal Aircraft Establishment, Farnborough, A. A. Griffith, was one of the first to explain the modern theory of the gas turbine and he researched the subject in detail in the 1920s. In 1928 an RAF officer cadet, Frank (later Sir Frank) Whittle, began to work on similar theories. In Germany Pabst von Ohain was working on the gas turbine principle in the 1930s, and it was his engine which powered the first turbojet powered aircraft to fly, the Heinkel He 178, on 27 August 1939. This single seat aircraft, with aluminium fuselage and wood wings, achieved 435 mph. In the malestrom of events following the outbreak of war, however, no further progress was made with the Heinkel, but Junkers and BMW did develop engines for a new generation of jet-propelled aircraft which were produced with astonishing speed by the Germans during the Second World War. Best known of these was the Messerschmitt Me 262, powered by two Junkers Jumo 004 engines. While the prototype was ready in 1942, it was not until late 1943 that production was sanctioned; when the Me 262 did appear it was sensational. It had a speed of 540 mph and an armament of four 30 mm guns. It was produced in fighter, bomber, and night fighter versions. Over 1,400 were built but by 1944–45 Germany was so hard pressed that only about 200 machines saw service.

Several other German jets were built, all showing great promise and far in advance of technological thinking anywhere else. They were too late to affect the course of the war, however, for by the time they appeared Germany's fate was sealed. The Arado 234 Blitz was a 461 mph jet bomber, the first in the world. It was in service late in 1944 but saw only limited use. A larger bomber was the Junkers Ju 287 which had six engines and forward-swept wings. In late 1944 hopes were being placed on a simpler jet fighter, the Heinkel 162A, sometimes called the 'Volksjäger', which was a simplified single jet type intended for mass production at 1,500 a month. Top speed was around 490 mph. Although the He 162 saw limited service, it appeared too late to swing the balance in the air war. In passing, the Germans also produced other ingenious designs not jet-propelled but rocket-propelled. There was the Messerschmitt Me 163 Komet fighter, a small dart-like machine which was shot skywards like a firework, to glide until its target approached, for its powered-flight duration was very limited. It was used against Allied bomber formations of Germany in 1945. The Germans also developed the first operational pilotless missiles, the V-1 and V-2 in 1944–45. While the A4 (V-2) was a straightforward rocket missile, the V-1 'flying bomb' was powered by an an Argus pulse-jet engine, a crude type of power unit acceptable only for this type of limited use.

It was the V-1 that Britain's first jet fighter was called upon to combat when it came into service in 1944. The first British jet aircraft was the Gloster-Whittle E.28/39, which flew on 15 May 1941. It was a tubby little single seater powered by the Whittle-designed turbojet. Basically the same engine was used in a twin-engined fighter, which first flew in March 1943 and entered service as the Meteor. With a speed of 493 mph, the Gloster Meteor was called upon to shoot down the V-1s which were difficult for piston-engined fighters to catch. The Meteor served in various models until the 1960s and beyond, and in 1946 held the World Speed Record of 616 mph. The second British jet fighter, the twin-boom de Havilland Vampire, was too late to see war service.

below, left
Twin-jet Gloster Meteor first flew in 1943, and entered RAF service in 1944, just before the Messerschmitt Me 262. It was later developed for many roles. This is a Mk 4, similar to the aircraft which set a world speed record in 1945

bottom
Modern air forces maintain a huge strategic transport capacity, able to airlift entire army divisions, their equipment and stores to any area of conflict. This is the capacious Douglas Globemaster of the 1950s

below
A Victor tanker aircraft refuelling a Buccaneer S.2 of the RAF

Meanwhile a Whittle engine had been taken to USA and powered the first American jet fighter, the Bell Airacomet which first flew in October 1942. The second US jet fighter and the first to see combat was the P-80 Lockheed Shooting Star, which first flew in January 1944. The Shooting Star and Meteor went on to give excellent service in the Korean War of 1950–53.

By the time of the Korean War, the Soviet Union had developed a fine swept-wing fighter, the MiG-15, first of a number of practical and rugged jet warplanes produced in great numbers by Russia and widely used in the 1950s, 1960s and 1970s.

During this period, there were many changes in combat aircraft, related mainly to technological advance. In the 30 years from 1945 aircraft became technically more complicated, speeds more than trebled and computers took over many of the functions that the pilot had previously worked out for himself. Speeds in excess of Mach I (the speed of sound) led to the swept

opposite, top
The Fiat G91 was a light tactical fighter built to a NATO requirement, used by the Italian and German air forces

opposite, centre
The North American F–86A Sabre was the MiG–15's main protagonist in Korea

opposite, lower
The classic Hawker Hunter. This is the Mk 4 of 1960

top
An Avro Vulcan heavy bomber of RAF Strike Command displays its capacious bomb bay

left
The enormous offensive load carried by this RAF Phantom typifies the punch a modern strike plane can deliver. Seven Cluster bombs, four Sparrow, and four Sidewinder missiles are shown. The Cluster bombs have enormous destructive potential, and are delivered from very low height against tactical military targets

161

opposite, top
French-built Dassault Mirage F1
of the Spanish Air Force, 1974.
A multi-role fighter type for
ground attack or interceptor use
it has a Mach 2·2 performance
at height and carries a variety
of alternative missile loads

opposite, lower
The SEPECAT—BAC Jaguar is a
successful Anglo-French attack
machine, serving both Britain
and France in the 1970s

above
The Panavia MRCA (multi-role
combat aircraft) is an attempt
to build a sophisticated attack
aircraft for Britain, Italy, and
Germany, for the 1980s.
Variable geometry wings and a
big offensive load which can be
delivered from a low height, are
characteristics

left
Neat formation flying in F—100
Super Sabres by the USAF
Aerobatic team, the Skyblazers

above
An F–105 in Vietnam in 1970

right
Hawker P.1127 was the first really successful vertical take-off machine in 1960

below
The Harrier can be used to give fleet cover and requires only small platforms—such as on the cruiser HMS *Blake* shown here—to operate. This dispenses with the need for long, large decks as required by conventional naval aircraft

left
The vertical take-off strike
aircraft points the way to future
development on naval airpower.
Here are Soviet VTOL Yak–36
fighters aboard the new carrier
Kiev in 1976. At left are
anti-submarine helicopters

above
Most successful of recent
tactical military transports is the
rear-loading Lockheed Hercules
which can operate from short
rough runways. It is in wide
service and is also used for
rescue roles

thin-section wings which are now taken for granted. Warplanes
became referred to as 'weapons systems'–they were designed
around their weapons, or specifically to deliver a type of
offensive load. Among the classic types which became known
(and in many cases used) throughout the world were the
Sabre, Hunter, Javelin, Phantom, Lightning, the French Mystère
series, the Thunderstreak, Thunderflash, Thunderchief, the
Canberra bomber, the Vulcan, Victor, Valiant, B-47, B-52,
Jaguar, MiG 21, and scores of others. Missiles became the enemy
of the combat aircraft, fired either from the ground or from
other aircraft. For a time it even seemed that missiles would
entirely replace guns as aircraft armament, but in recent years
the value of the gun has been re-discovered and few fighters
of modern times rely solely on missile armament. Long-held
theories have been turned upside down from time to time; for
example in the Vietnam War of the 1969–73 period, the
vulnerability of strategic bombers to fairly simple surface-to-air
missile systems was demonstrated, and bombing raids over North
Vietnam became surprisingly costly in lost aircraft and crews.
The value of the 'slow' aircraft was discovered again in wars
against unsophisticated enemies, as in Vietnam. All the most
expensive and the most modern hardware would not flush out
hidden guerilla troops half as effectively as old Dakotas fitted
with Gatling guns, or COIN (counter-insurgency) aircraft of the
piston engine type like the Skyraider.

As warplanes got more expensive there was a move to simpler,
cheaper, less sophisticated types. There thus came an age of

the lightweight fighter–the Gnat, the F-5 Freedom Fighter, and
the F-16 are classic examples. By the late 1960s, vertical take-off
and landing added a new dimension with the Harrier
fighter-bomber which could operate from jungle clearings and
liberated the aircraft at last from the need for long and
vulnerable runways. VSTOL aircraft also make sense at sea
and arrived on the scene, none too soon. For aircraft carriers
able to operate big aircraft were getting too expensive and too
big. But a VSTOL fighter could operate from a small simple
ship, and sooner rather than later this could make the
conventional aircraft carrier a thing of the past, to disappear
into limbo to join the battleship as an obsolete type.

The Age of the Airlines

The limitation on airline development in the 1930s was one of range and endurance. Airline routes had been set up *across* continents–for example America and Europe–but the next quest for achievement was to *link* continents by air as an everyday routine. Special flights, like Lindbergh's or Alcock and Brown's, linked continents quite early on in the time scale, but airline progress depended very much on the development of aeronautical technology. It was 1928 before the first commercial aircraft flew the Atlantic east-to-west. This was a Junkers W.33 cargo monoplane, the 'Bremen', and it was a specially prepared flight.

Another Junkers design pointed well ahead to the future. This was the Junkers G.38 giant airliner, the 'Jumbo' of its day, which first flew in prototype form in 1929. The two aircraft built, named 'Von Hindenburg' and 'Deutschland', were flagships of the Lufthansa line and served on principal routes from Berlin from 1931 until the Second World War started. With four Jumo 750 hp diesel motors, the G.38 had a 2,150-mile range and a 127 mph cruising speed. Span was $144\frac{1}{3}$ ft, length 76 ft, and the massive wings had a $32\frac{3}{4}$ ft chord and were 5 ft 7 in thick at the roots, allowing access to the engines and passenger saloons inboard of the inner engines! The G.38 had a crew of seven and carried 34 passengers–a large payload for 1931.

Initial ideas for spanning the oceans were linked to the potential of the airship. Airships had been used quite extensively in the First World War by both the Germans and the British. The British in particular used them for ocean reconnaissance, and this led to the idea of using very large airships to link the British Empire by air. The 1920s saw much activity to achieve this aim, when Britain built two giant airships, the R.100 and the R.101, for the purpose. Even before they were completed,

however, the Germans stole a march on the British by making the first commercial trans-Atlantic flight by airship with the giant 'Graf Zeppelin' in 1928. In 1929 the 'Graf Zeppelin' circled the world in three weeks and this appeared to justify the faith put into the future of big airships. In October 1930 the R.101 left her base at Cardington, Bedfordshire, for her maiden flight to India. Her passengers for this first prestigious journey included the Air Minister, Lord Thomson, and Sir Sefton Branckner, Director of Civil Aviation. North of Paris she crashed in flames in a rainstorm, a victim of the very unstable and explosive hydrogen gas then used as the filler for lighter-than-air craft. Only six of the 54 passengers and crew aboard the luxurious airship survived.

The disaster brought commercial airship development in

above, left
The US Navy's *Shenandoah* was the first airship to be raised by helium gas, in 1923. Helium is an inert gas which was, of course, much safer for airship use than hydrogen

above, right
In the small Italian semi-rigid airship N.1, later 'Norge', the explorer Amundsen pioneered flying the North Polar route in 1926, only just beaten by Richard Byrd in an aircraft

right
The disastrous accident to the 'Hindenburg' in May 1937 ended passenger carrying airship activity

opposite, top
The Short SD 330, a utility transport and feeder liner of the 1970s, and a successor to the earlier Skyvan

opposite, lower
Another utility liner and feeder transport which has enjoyed world wide success is the Britten-Norman Islander, a 170 mph machine carrying up to nine passengers

Britain to an untimely end. The Germans were more optimistic, however, and took the chance of establishing a regular trans-Atlantic service in 1936 with the fine new 'Hindenburg', which was bigger than any previous craft. In May 1937, however, came a spectacular disaster when 'Hindenburg' crashed in flames while coming to her mooring mast at Lakehurst, New Jersey. Some 35 passengers and crew were killed and this accident led to the complete demise of the giant airship. Over the years other accidents, especially to US airships (the 'Shenandoah', 'Akron' and 'Macon', all US Navy craft) had discouraged further development, at least of large, rigid (ie framed) airships. Thereafter airship development was restricted to small, non-rigid craft; the American Goodyear firm was the specialist builder and the US Navy a prime user in the Second World War and after. Fitted with inert helium gas, these were safe, popular and economical craft, ideal for ocean anti-submarine patrols of long duration. Post-war civil Goodyear blimps were used for advertising and survey work. In the 1970s some remained in commercial use, a popular modern employment in this technological age being as an aerial TV camera platform for outdoor events–an interesting combination of a modern electronic science with one of the oldest and basic forms of flying.

The alternative in the 1920s for trans-oceanic work appeared to be the flying boat, partly because of the theory that, in the event of emergency, it could put down on the ocean. The Dornier Do X was the first manifestation of the idea, a huge 12-engine

above
Short 'Empire' flying boat,
'Caledonia' of Imperial Airways,
over New York after a
trans-Atlantic proving flight in
1939.

right
The Short Kent was a refined
flying boat for Imperial Airways
in 1931. Note the enclosed
cockpit

below
The Short Solent was a
post-war passenger
development from the
Sunderland patrol flying boat,
last of the flying boat airliners
used by Britain's national airline
in the 1950s

craft carrying up to 100 passengers. It was way ahead of its time, and the biggest aircraft of its day. It was magnificent, but a technical failure due largely to the inadequacies of the available engines. During a world tour it was damaged and this ended the project.

In Britain the Short company built some fine, large, biplane flying boats. A passenger type, the Calcutta, was used by Imperial Airways for the Mediterranean stage of the Britain-India route. A larger machine for the RAF, the Singapore, was used by Sir Alan Cobham in 1928 to pioneer a 23,000-mile route around Africa to show how passenger services could be flown to serve that continent from Britain. (Cobham subsequently did much in the 1930s to pioneer in-flight refuelling to achieve long range–substantially the same techniques as are still in use for warplanes today.)

Largely as a result of Cobham's trail-blazing, Imperial Airways started an Empire Air Mail scheme in 1937, the all-metal, four-engined flying boats used being ordered in 1935. Eventually, 42 Short 'Empire' or 'C' class boats were built. These were magnificent machines used on services to Australia and South Africa, and some were supplied to the Australian airline, Qantas, in 1938 for a reciprocal service. Only 17-24 passengers could be carried, however, as airmail took up a large part of the payload. The 'Empire' design was also developed into the Sunderland long-range aircraft used by the RAF for convoy patrol in the Second World War. A further development was the Short-Mayo Composite designed by Major Mayo, who had advised on the 'Empire' design. The Composite exploited the pick-a-back idea, the big flying boat taking up a small seaplane which was then launched to complete the journey. Using this technique, a trans-Atlantic flight was made from Ireland to Montreal in July 1938 and another flight was made from Dundee to the Orange River in South Africa in October 1938.

above
Short Mayo Composite. A Short 'Empire' flying boat with Mercury seaplane in 1938

below
The de Havilland Heron was a widely used feeder liner in the 1950s and 1960s. This Heron is a BEA flying ambulance serving the Scottish Islands

In the meantime, the up-and-coming Pan-American Airways
line had ideas for flying routes across the Pacific. There was
already a route down the east coast of South America flown
by the New York, Rio, and Buenos Aires line, a company which
was absorbed by Pan Am in 1931, putting them into the flying
boat business with the superb, high-wing Consolidated
Commodore, a beautiful luxury liner used on the route. Pan
Am asked the United Aircraft firm to build bigger flying boats,
and this resulted in the magnificent S-40 and S-42 models,
designed by the erstwhile helicopter pioneer, Igor Sikorsky. The
S-40 was used to expand the Latin American operations, while
the S-42 'Flying Clipper' opened up the trans-Pacific route in
1934. San Francisco, Hawaii, Wake Island, Guam, Manila, and
Hong Kong formed the route in 1935, and two years later another
route was opened to New Zealand. The Martin M-130 flying
boat also appeared in 1935, especially to work the trans-Pacific
routes.

Boeing also built a superb new flying boat for Pan Am, the
Model 314, for the first trans-Atlantic service, the greatest
commercial prospect of all. The Boeing 314 carried 40 passengers,
had four 1,500 hp Wright Cyclone engines, a 183 mph cruising
speed, and a 3,500-mile range. The first ocean flight took place
between Port Washington and Lisbon in June 1939, an immense
success that was cut short by the outbreak of war, though the
314s survived to post-war days.

Lufthansa also had eyes on the trans-Atlantic route and their
method in 1937 was to use large floatplanes launched by

above
In 1937 Lufthansa made a series of flights between the Azores and New York using a Blohm & Voss Ha 139

right
Lufthansa pioneered air mail services to South America in 1934 using Dornier Wal seaplanes. The aircraft were lifted onto a mid-ocean base ship for servicing and refuelling, then relaunched by catapult

below
In August 1938 Lufthansa flew the Berlin–New York route with this Focke-Wulf Fw 200 Condor. Flying time was 24 hours 57 minutes

catapults from specially-built carrier ships. The Blohm und Voss company built the Ha 139 for this task, a clean all-metal design with four Jumo 600 hp diesel motors giving a top speed of 196 mph. Span was $88\frac{1}{2}$ ft and length 64 ft. These were mail planes, with four-man crews and carrying 1,058 lb of mail. The special ships, *Friesenland* and *Schwabenland*, took station in the Atlantic to refuel the aircraft, carry out servicing and relaunching them on their way. Seven return journeys were made to New York from Horta in the fall of 1937 to test the feasibility of the project. In 1938–39 a full service was started, with 40 crossings on the North Atlantic route and 60 on a South Atlantic route–Bathurst-Natal (in Brazil)-Recife–then the outbreak of war ended the service. By then, however, a fine, big, six-engined flying boat was on order from Blohm und Voss, the Bv 222 Wiking. This was to carry 16-24 passengers at over 200 mph and was clearly inspired by the British Short 'Empire' boats. However, with the advent of war, this Lufthansa scheme came to nought, the protypes seeing only military service.

Boeing now showed the way ahead with a design to meet the future requirements of US airlines. This was the four-engine Model 307 Stratoliner, the world's first fully-pressurised airliner, able to fly high 'above the weather' with increased economy and enhanced passenger comfort. The Stratoliner was based on the B-17 bomber as far as wings, engine and tail were concerned, and it was in production in 1937–38 for Pan Am and TWA. All the machines were taken into military service in the Second World War, and the design proved astonishingly durable—surviving aircraft were still flying in the 1960s, looking so modern that a layman would have difficulty in appreciating their historical significance. The Stratoliner carried 33 seated or 20 sleeper passengers.

below
Canadair North Star was the Canadian-built version of the Douglas DC–4, having Rolls-Royce Merlin engines instead of Pratt & Whitney R–2000s. This veteran was in service with the Canadian Government until 1976, having first flown in 1947

bottom
The BAC 1–11 was a short-haul rear-engined jet which has seen wide and successful service since the early 1960s. In the background are a Vickers VC10 and a Boeing 707

above
The highly successful Boeing
707 has been one of the most
widely used airliners since it
entered service in 1958. Service
vehicles here cluster round a
Lufthansa machine at Frankfurt

right
Smallest of the Boeing jet liner
family is the twin-engine 737
'City Jet' for short haul work.
Lufthansa is a major user of this
type on its European routes

top
The original de Havilland
Comet I of 1952

above
Tupolev Tu–104 was Russia's
first jet liner in 1956. It carries
70–100 passengers at 497 mph,
and has two wing root engines.
Unusually among civil aircraft,
braking parachutes were used
on the Tu–104 in difficult
landing conditions

left
The original DC–8, here
operated by Air Canada

below
Sud-Aviation Caravelle of 1955
started the trend towards the
rear engine layout for airliners

Seen here in the colours of the American line Braniff, the Boeing 727 is the most successful of the tri-jet medium range airliners

The Second World War stopped all but 'official' passenger and freight flying and many airliners were impressed for military service. Such types as the DC-3 earned extra fame as paratroop and supply droppers under the service designation C-47, or name Dakota. The DC-4, in effect a four-engined, enlarged version of the DC-3, appeared as a prototype in 1938 and saw extensive military service as the C-54 Skymaster. It had a comfortable trans-Atlantic capability and spelled the death knell for flying boats on the transoceanic routes. The first landplane airliner over the Atlantic, however, had been a Focke-Wulf Fw 200 Condor, a superb, streamlined, 26-passenger plane. In August 1938 this four-engined, 226 mph machine flew Berlin-New York, and return, in less than 25 hours westbound and 20 hours eastbound. Late that year the Condor flew Berlin-Tokyo with only three refuelling stops in 46 hours. The Condor ended its days as a Luftwaffe long-range bomber harassing convoys.

The DC-4 and the enlarged DC-6, which was pressurised, saw

left
Ilyushin Il–76 (foreground)
was designed as a military
transport, while the Il–62
(background) is a long-range
airliner similar in configuration
to the British VC–10

below
A Tupolev Tu–144 of Aeroflot,
in 1975

Boeing 737

Boeing 727 B

Douglas DC-3 (1955–60)

Convair CV-440 „Metropolitan" (1955–68)

Boeing 707

Vickers V-814 „Viscount" (1958–71)

Airbus A 300

McDonnell Douglas DC-10

Lockheed L 1049 G „Super Constellation" (1955–67)

Boeing 747

above, left
Lufthansa's propeller aircraft of
the 1950s (drawn to a common
scale) . . .

above, right
. . . and the 1970s generation
of jet liners operated by the
same company (also to a
common scale)

below
A specialized cargo carrier with
opening nose doors which set
a trend for later larger types was
the successful Bristol Freighter
and Super Freighter (shown
here) of the 1950s

extensive post-war service; about 400 of the latter were built,
and many were still in use on charter routes in the 1970s. Last
of the piston-engined Douglas airliners was the DC-7, the 7B
and 7C, the latter serving with nearly all the major international
airlines in the late 1950s before the jet liners replaced them.
The main contemporaries on transoceanic routes were the
Boeing Stratocruiser and Lockheed Constellation. The former
was a commercial airliner development of the B-29 and B-50
bombers, a two-decker which carried up to 100 passengers. Most
graceful of this airliner generation, however, was the Lockheed
Constellation and a lengthened development, the Super
Constellation. Although conceived for TWA in 1939, the 'Connie'
went straight into military service as a transport before starting
a long, post-war career with most of the major airlines. Triple
tailfins and a lean, aerodynamic fuselage were its distinctive
features.

For shorter internal routes there were now new contenders
for the market filled by the DC-3. Outstanding was the Convair
240 and 440, which saw service with major world and internal
US airlines. The Lockheed Electra, with four turboprops, was
a later contender, taking cruising speed to over 400 mph. Most
successful of all, however, was the superb Vickers Viscount,
Britain's finest commercial success since 1945. More than 440
Viscounts were sold to well over 60 airlines, and the aircraft
achieved the considerable feat of breaking into the American
internal airline market. Like other successful airliners in their
day, the Viscount happened to be the right aircraft at the right
time and the Rolls-Royce Dart propjets made it a smooth and
economical machine to operate–and it was first in its field, as
a prototype in 1948 and in service in 1950. Many still operated
in the 1970s, and it remains one of the world's classic aircraft.

above
A pair of Canadian STOL utility transports in United Nations service in the Middle East—a twin-engined de Havilland Caribou and a single-engined de Havilland Otter

left
A DC–10 of Lufthansa

below
Lockheed Tristar was the rival to the DC–10 and A.300 Airbus for the wide-bodied jet market of the 1970s

above
The beautiful Lockheed Super Constellation, in BOAC service

right
The Boeing Stratocruiser, here in service with BOAC

below
One of the most successful contenders as a 'Dakota' replacement has been the Fokker F.27 Friendship (also built by Fairchild in USA), which has been sold throughout the world. It is powered by Rolls-Royce Dart propjets, cruises at 295 mph, and carries up to 50 passengers

The jet age started for airline traffic in May 1952 when the beautiful de Havilland Comet made the historic, first BOAC commercial jet flight from London to Johannesburg. Soon, Comets were also flying to Ceylon, Singapore and Tokyo and some foreign airlines placed orders. But disastrous crashes in 1954, which were not immediately explicable, caused the type to be taken out of service. Though a later Comet 4 was developed and proved successful, Britain's lead in this field was lost, for the major American builders, Boeing and Douglas, gained time to evolve and build their own ideas of jet liners. These ideas emanated in the celebrated Boeing 707 and the Douglas DC-8 (the Convair 990 was less successful). In service in 1958 these machines, with four underslung engines, were capable of much further development, and the 707 and DC-8, perhaps more than any other airliners, have 'shrunk' the world so that, today, travel by air, rather than by ship, is the most common method of transoceanic travel. Boeing and Douglas then both developed 'families' of smaller and larger airliners derived from the basic concept—the Boeing 737, 727 and 747 Jumbo, and the Douglas DC-9, DC-10 and so on, becoming very familiar types.

It was the French, however, who first pioneered the classic rear-engine layout which was to be followed by many American and British airliners. In 1955 the Sud-Aviation firm produced the Caravelle, which was sensationally successful. It followed the lines and layout of the Comet but instead of having the engines buried rather inaccessibly in the wing roots, they were set at the tail, easy to service and replace. Similar layouts were adopted for such types as the DC-9, Boeing 727, Trident, VC-10, BAC 1-11, and Tupolev 134, as well as for smaller, executive aircraft.

The Soviet Union, with a huge market in the Soviet bloc countries and further abroad, also became very large builders of airliners from the 1950s onwards. The Il-12, a short-haul type like the DC-3 (which was licence built in Russia), has been widely used. The Tupolev 104 was Russia's first jet liner in 1956, being derived from the Tu-16 bomber. The Ilyushin 18 and Tupolev 114 were successful, large propjet machines, but in more recent years the American and British influence has shown in types like the Tu-154, very similar to the Boeing 727, and the Il-62, like the BAC VC-10.

Hawker Siddeley Trident III is typical of the 1960s generation of tri-jets for short and medium range passenger work. It cruises at about 600 mph with about 125 passengers. This one was in BEA service in 1972

above
An–24 is a short-haul turboprop airliner in the same category as the Fokker Friendship

opposite, bottom
VFW-Fokker 614 is a German-Dutch attempt of the 1970s at a 'Dakota replacement'. This short-haul twin jet has unconventional engine placing to simplify construction and maintenance. It carries up to 44 passengers at about 450 mph

185

The 1970s were really the age of the 'Jumbo' jet, however. The quest for economy and scale of turn round means that operating costs must be optimised—and that means carrying as many passengers per flight as possible. Boeing pioneered this idea in the 1960s when they studied forward projections of passenger traffic. They evolved the 747, an immense machine, span $195\frac{1}{2}$ ft, length $231\frac{1}{3}$ ft, height $63\frac{1}{2}$ ft, and with a gross weight of 710,000 lb (over 300 tons). The range (varying with model) is about 5,500 miles and the cruising speed 585 mph. From 374 to 490 passengers—depending on seating arrangements—and an 18-wheel landing gear started to turn the Jumbo jet project into an exercise in impressive statistics. But they all added up to profitable operation. Pan Am were the first Boeing 747 customers, in 1969, and since then nearly all major Western airlines have become users. The passengers, mostly, like Jumbo jets, too, for compared with earlier 'slim' aircraft—even 707s—the Jumbo jet is roomy and comfortable. Douglas, with the DC-10, and Lockheed, with the Tristar, also entered the Jumbo market, initially for short- to medium-range work, while a European consortium also built a very successful A-300 Airbus. These aircraft carry over 300 passengers, the first two with engine aft configuration, the Airbus with underwing

top
First flown in 1947, the Antonov An–2 was one of the last biplanes in large scale production. A tough utility machine it has seen wide service, well over 5,000 being built. As a utility transport it carries up to ten passengers

centre
The Antonov An–22 is one of a family of huge Russian rear-loading cargo and passenger carriers which while seeing extensive commercial service also has a military role

engines. Long-range projects of all these were planned in the 1970s.

Subsonic jets of huge capacity made economic sense. More controversial, however, were supersonic jets such as the BAC-Aerospatiale Concorde and its Russian rival, the Tu-144. Boeing in America abandoned an SST (supersonic transport), the Boeing 2707-300, in 1971, mainly on the grounds of immense cost and the prospect of little return. Britain and France jointly developed Concorde over a 15-year period and commercial flights started in 1976–to some on a note of optimism for the future while others were very pessimistic. This pessimism may not augur badly for the future–the Wrights, Langley and all the other pioneers realised their plans in the face of considerable criticism. 'It will never fly' is perhaps the most frequent assertion that has been made in about 100 years of aviation history!

above
First commercial flight of the Concorde from London's Heathrow airport, Bahrain bound on 21 January 1976. Note how the nose is lowered for improved visibility for take-off and landing

opposite, bottom
In 1961, the Tupolev Tu–114 was the world's largest airliner. A propjet, it was derived from the Tu–20 bomber, carried up to 220 passengers at about 480 mph, and has been used on long-distance routes

right
A balloon race across the peaceful English countryside in 1975 — the hot-air balloon proved a useful vehicle for displaying advertisements as is evident here

below
The Czech Zlin Trener is one of the specialist types produced in the 1960s for aerobatic work. It was widely used in aerobatic contests, having many of the characteristics of the aircraft of the 'golden days' of flying

bottom
A pedal-powered aircraft in flight. This one was built and flown by the Aeronautics Department of Southampton University

Epilogue

While the aviation world is generally thought of in the late 1970s to be a highly technical mix of missile-armed attack planes, costly SSTs, and colourful Jumbo jets, there has never been a time in aviation history when the ordinary enthusiast can get so close to 'living' the past as the pioneers experienced it. Nobody living today was around in the days of the Montgolfiers, yet hot-air balloons not dissimilar to their originals exist in greater numbers than at any time in the past. Balloon meets where hot-air balloons can be seen in profusion are not uncommon. The budding follower of Lilenthial, Chanute, or Pilcher can equip himself with a hang glider which any of those pioneers would have recognised and flown at once. Even the humble kite as flown by such as Cayley or Hargraves has come back into its own, with kite-flying becoming a national craze in Britain complete with a Kite Flying Association! For the more determined aviator there has been a revival in the building of ultra-light aircraft, just as there was in the 1920s, for conventional light aircraft made by the big manufacturers are beyond the pocket of most. And some determined men are still pedalling away at man-powered flight, with a £50,000 prize still waiting to be won for anyone able to fly round a half-mile course.

Index

Ader, C., 21, 32
Aeronca C-3, 92, 95
  Sedan, 91
Aerospatiale Puma, 115, 116
Aichi D3A, 153
Air France, 63
Air Transport and Travel, 62
Airbus, 186
aircraft carriers, 149
airlines, early, 60 et seq
airships, early, 22 et seq
  passenger, 168, 169
Airspeed Horsa, 155
Albatros fighters, 59
Alcock, J., 66
Antonov ANT-6, 130, 131
  An-2, 188
  An-22, 188
  An-24, 185
Arado Ar 234, 158
Armstrong Whitworth Argosy,
  65, 70
  Atalanta, 71
  Siskin, 119
Arrow Active, 96
Atlantic crossings, early, 65
Auster, 94
  Aiglet, 85
Avro, 50
  Baby, 85
  Lancaster, 141, 144 et seq
  Manchester, 142, 143
  Rota, 103
  Tandem, 60
  triplane, 35
  504K, 51, 80; 504N, 80
  506, 81
  Vulcan, 161

balloons, 10 et seq, 190
'barn storming', 81
Batten, J., 87
Beech Beechcraft 17R, 94
  Duke, 87
  18, 78
Bell, A.G., 33
Bell helicopters, 112, 113
  P-39 Airacobra, 156
  P-59 Airacomet, 161
BE2, 53
Benoist flying boat, 60
Blackburn Buccaneer,
  see Hawker Siddeley
Blanchard, J. P., 12, 15
Blériot, L., 34
Blériot XI, 31
Bloch 220, 78
Blohm & Voss Ha 139, 176
Boeing, B1, 62
  B9, 73
  B-17 Flying Fortress, 139 et seq
  B-29 Superfortress, 145, 148
  B-47A, 151
  B-52H Stratfortress, 151
  Monomail, 73
  MB-3A, 122
  P-12 (F4B-1), 123
  P-26, 124, 126
  Sea Knight, 114

Stratocruiser, 182, 184
40A, 73
80A, 73
247, 73 et seq
307, 177
314, 174
707, 178, 185
727, 180
737, 178
747, 186
2707-300 project, 189
Boeing-Stearman Kaydet, 123
de Bothezab, G., 101
Boulton Paul Defiant, 134
  Overstrand, 122
Bréguet, 60
  Gyroplane, 100
  Gyroplane Labaratoire, 103
  helicopter, 100
  14, 61
Brennan, L., 102
le Bris, J-M, 20
Bristol, 50
  Beaufighter, 134, 138
  Belvedere, 115
  Blenheim, 130
  Boxkite, 35
  Bulldog, 121
  Fighter, 46, 120
  Freighter, 182
British Aircraft Corporation
  Concorde, 186, 187, 189
  Jaguar, 162
  Lightning, 154
  1-11, 177
  VC-10, 175
Britten-Norman Islander, 171
Brown, A. W.-, 66
Bucker Jungmeister, 127

Camm, Sir Sidney, 128
Canadair North Star, 177
Caproni Ca, 42
Caudron G23, 62
  G111, 39
  Luciole, 88
  Phalene, 88
Cayley, Sir George, 17, 19, 22, 99
Cessna Skylane, 94
  C-37, 93
  Skymaster, 90
Chambers, W. I., 38
Chanute, O., 21
Charles, J. A. C., 14
de la Cierva, J., 102
Cobham, Sir Alan, 81
Cody, S. F., 36, 37
Consolidated B-24 Liberator,
  140, 141
  Catalina, 149
Curtiss, G. H., 35, 36
Curtiss Condor, 73
  Hawk, 122
  JN-2 'Jenny', 59, 80
  NC flying boats, 65, 66
  P-40, 135, 137
  PN-8, 122
  Pusher, 41
  Racers, 95

Curton, G., 29

Daimler Hire, 62
Dassault Mirage, 162, 166
de Havilland, G., 81
de Havilland, 35
  Comet, 179
  D.H.1, 52
  D.H.2, 53
  D.H.4, 63, 80
  D.H.9, 62, 67, 119
  D.H.51, 79
  D.H.53, 81, 82
  D.H.54, 70
  D.H.60 Moth, 82, 83
  D.H.66, 70
  D.H.71 Tiger Moth
    (monoplane), 83
  D.H.80 Puss Moth, 83
  D.H.82 Tiger Moth, 82, 83
  D.H.83 Fox Moth, 83, 85
  D.H.84 Dragon, 83
  D.H.85 Leopard Moth, 83
  D.H.87 Hornet Moth, 83
  D.H.89 Dragon, 83
  D.H.90 Dragonfly, 83
  D.H.94 Moth Minor, 83
  Gipsy Moth, 83
  Heron, 173
  Mosquito, 135, 136
  Vampire, 158
  see also Hawker Siddeley
de Havilland Canada Otter, 183
Deperdussin-Bechereau, 49, 50
Dewoitine D26, 125
  D27, 125
  D510, 126
  332, 78
von Doblhoff, F., 105
Doolittle, J., 75, 96
Dornier Do 17, 130
  Do X, 170
  Wal, 176
Douglas A-20 (Boston), 136, 142
  A-24 Dauntless, 152
  C-124 Globemaster, 159
  DC-2, 76
  DC-3, 75 et seq, 182; Dakota,
    165
  DC-4, 180
  DC-8, 179, 185
  DC-10, 183, 186
  TBD-1 Devastator, 152
  see also McDonnell Douglas

Ellehammer, J., 101
Ely, E., 38, 39
English Electric, see British
  Aircraft Corporation
Esnault-Pelterie, R., 34

Fairey Albacore, 147
  Battle, 134
  Rotodyne, 116
  Swordfish, 147, 150
Farman, H., 33
Farman F.303, 63
  Goliath, 62
  H20, 50

Shorthorn, 51
  Voisin, 33
FE2, 51
Fiat CR32, 126
  CR42, 123
  G18, 78
  G91, 160
Flettner helicopters, 106
'Flying Flea', 88
Focke-Achgelis, 104, 177
Focke Wulf Fw 34, 126
  Fw 190, 139, 140
  Fw 200, 129, 176, 180
Fokker, A., 50, 53, 69
Fokker DVII, 58
  E111, 39, 53
  F111, 67, 69
  FVII, 69
  FXII, 71
  F-10, 69
  F-27, 184
  Triplane, 55
  -VFW-Fokker 614, 185
Ford Trimotor, 69, 70
Friedrichshafen G111, 59

Garros, R., 53
gas turbines, early, 157 et seq
General Dynamics B-58, 167
Giffard, H., 22
gliders, 96 et seq
  military, 155
Gloster E28/39, 158
  Gauntlet, 122
  Gladiator, 121
  Grebe, 119
  Meteor, 159
Gnome engine, 33
Goodyear dirigibles, 27, 28, 168
Gordon Bennett races, 49, 96
Gotha bombers, 59
Grahame-White, C., 49, 60
Grain Kitten, 81
Grieve, M., 66
Grumman F4F Wildcat, 152
  F6F Hellcat, 153
  F-14 Tomcat, 157
  TBF Avenger, 152
de Gusmao, L., 11

Handley Page, 50
  H.P. 42, 71, 72
  H.P. 45, 71
  0/100, 56
  0/400, 57, 64
  Victor, 159
Handley Page Air Transport,
  62, 64
Hanriot HD-1, 59
Hargrave, L., 22, 23
Hawker, H., 66
Hawker/Hawker Siddeley
  Argosy, 168
  Buccanneer, 159
  Cygnet, 81
  Fury, 121
  Harrier, 164, 167
  Hart, 122
  Hunter, 161

Hurricane, 125, 128, 129, 132
Nimrod, 154
P1137, 164
Tempest, 155
Trident, 185
Typhoon, 155
Heinkel He 51, 126
He 70, 78
He 111, 78, 129
He 162, 150, 158
He 178, 158
helicopters, 99 *et seq*
Henson, W. S., 17, 19
Howard DGA-2, 95
Huey Cobra, 113
Hughes, 96

Ilyushin Il-2 Stormovik, 156
Il-12, 185
Il-62, 181
Il-76, 181
Imperial Airways, 62
Instone Airlines, 62

Jefferies, J., 12, 15
jet engines, *see* gas turbine
Johnson, A., 86
Jullien, P., 22
Junkers F13, 68
G31, 71
G38, 168
J4, 59
J10, 59
Ju 52, 71, 73, 131
Ju 86, 78
Ju 87, 133
Ju 88, 133
Ju 287, 158

Kellett KD-1, 103
Keystone B-6, 124
KLM, 62
Krebs, A., 23

de Lana, F., 9
Langley, S. P., 28
Latécoire Laté-28, 63
Latham, H., 35
Learjet 35/36, 168
light aircraft, 79 *et seq*
Lilienthal, O., 21
Lindbergh, Col. C., 89
Lioré et Olivier LeO 213, 63
Lockheed Cheyenne, 113
Constellation, 182, 184
C-130 Hercules, 165
Electra, 77, 78
Electra turboprop, 182
Hudson, 136, 139
Orion, 77
P-38 Lightning, 137, 138
P-80 Shooting Star, 161
Tristar, 183, 186
Vega, 75, 96
12, 78
Luardi, V., 15
LVG, 50, 51
C VI, 46

man-powered flight, 191
Martin B10, 125
NBS-1, 62
Maxim, Sir Hiram, 21
May, T., 21
Mayfly, 44
McDonnell Douglas F-4
Phantom, 6, 156, 161
XV-1, 113
see also Douglas
McIntosh, J. C., 67
Messerschmitt Me 109, 127, 128
Me 110, 132
Me 163, 158
Me 262, 158
Mignet, H., 88
Mikoyan Mig-15, 158
Mig-23, 157
Mil Mi-6, 117
Mi-10, 118
Mi-12, 118
Miles light aircraft, 85
Master, 86
Mitchell, R. J., 129
Mitsubishi A6M Zero, 152, 153
Mollinson, J., 87
Montgolfier balloons, 10 *et seq*
Morane Saulnier, 34
light aircraft, 89
Type L, 52

Nakajima B5N, 153
naval aviation, 149 *et seq*
Nieuport, 36
Scout, 51
62-C1, 126
North American P-51 Mustang, 136, 137, 138
F-86 Sabre, 160
F-100 Super Sabres, 163
T-6 Harvard, 139
Northrop F-5 Freedom Fighter, 167
Oemichen, E., 101

Panavia MRCA, 163
Parnell Pixie, 81
Parer, R., 67
de la Pauze, P., 100
Pescara, the Marquess Roul, 101
Pescara helicopter, 100
Percival Gull, 86
Proctor, 86
Phillips, H., 30
Phillips, W. H., 100
Piasecki, F., 115
Pilatus Turbo Porter, 175
Pilcher, P., 21, 22
Piper, 92, 93
Caribbean, 93
Cherokee, 91
Cruiser, 93
Cub, 93
Pacer, 93
Tri-pacer, 93
Porte, J., 65
Potez 9, 60
light aircraft, 89
Puliyzer races, 95

Radar, 134
Rallye, 84
Rata I-16, 131
RE8, 51
Reitsch, H., 104
Renard, C., 23
Republic P-47 Thunderbolt, 137
F-105 Thunderchief, 163, 164
Robert, M. -N., 14
Robin, 84
Rockwell B-1, 166
Rogers, C., 49
de Rozier, J. F., 13, 15
Rumpler Taube, 50
Ryan 'Spirit of St Louis', 92

St Petersburg-Tampa Air Boat Line, 60
Salmson Cri-Cri, 89
Santos-Dumont, A., 23, 24, 31, 32
Sarti, V., 99
Saunder-Roe Skeeter, 115
Savioa-Marchetti SM79, 126
Schneider Trophy, 50
SE5, 56
SEPECAT Jaguar, 162
Short Calcutta, 173
Empire (C) class, 172, 173;
Mayo Composite, 173
Kent, 172
Scylla, 71, 72
Silver Streak, 69
Solent, 172
Stirling, 143
Sunderland, 152
S.27, 44
S.41, 40
SD 330, 171
184, 48
Sikorsky, I., 50, 100 *et seq*
Ilya Mourometz, 45, 50
Sikorsky R-4, 104, 105
S-40, 174
S-51, 109
S-55, 109
S-56, 109
S-61, 110, 111
VS-300, 104, 105
Slingsby Eagle, 98
Smith, K., 66
Smith, R., 66
Sopwith, T. O. M., 50
Sopwith Camel, 53
Pup, 49, 53, 58
Tabloid, 50
Triplane, 52
Spad VII, 54
XIII, 55
Sperry, E., 75
Stampe SV4, 88, 96
Stinson Detroiter, 73
Reliant, 92
SM-1, 92
Stout, W. B., 69
Stringfellow, J., 19
Sud Aviation Alouette, 117
Caravelle, 179, 185
Concorde, 186, 187, 189
Djinn, 110, 117

Frelon, 117
Sueter, Capt. M., 41
Supermarine Spitfire, 125, 129, 132
S6B, 128
Walrus, 148

Taylorcraft Model A Cub, 92
du Temple, F., 20
Thomas-Morse Scout, 59
transport services, early, 60 *et seq*
Trenchard, H., 57
Tupolev Tu-104, 179, 185
Tu-114, 189
Tu-134, 117
Tu-144, 181

Vedrines, J., 49
Verdon-Roe, A., 35
Vertol, 115
VFW-Fokker 614, 185
Vickers Gunbus, 52
Valentine, 124
Vimy, 66, 119
Viscount, 175, 182
Wellington, 129 *et seq*
see also British Aircraft Corporation
da Vinci, L., 8, 9, 99
Voisin, G., 31
Voisin Boxkite, 31, 32
Vought F4U Corsair, 153

Waco Hadrian, 155
Weir helicopters, 105
Westland Dragonfly, 182
Lynx, 115
Puma, 115, 116
Scout, 115, 116
Wessex, 111, 117
Whirlwind, 110
Wibault 282 TI2, 63
Wolfert, Dr K., 23
Wright, O., 29 *et seq*
Wright, W., 29 *et seq*

Yakovlev Yak 9, 131
Yak 24, 118
Yak 36, 165
Yokosuka D4Y Suisei, 153

von Zeppelin, Count Ferdinand, 25 *et seq*
Zeppelins, 25 *et seq*, 41, 57
early passenger flights, 60
'Graf Zeppelin', 168, 169
Zlin Trener, 190